The Book of Numbers

Images of the World's Ancient and Modern Numerals with Explanations

古今世界各国数字及图文释义

John Oxenham Goodman

约翰• 奥克森那姆• 古德曼

Sydney, 2017

Other Books by John Oxenham Goodman

约翰•奥克森那姆•古德曼

Volume 1—The Chemical Elements and the 88 Constellations in Art, Myth and History 历史神话艺术中的化学元素和 88 星座。Amazon, 2015. ISBN-10: 1519504756; ISBN-13: 978-1519504753

Volume 2—The Sun, Planets, Dwarf Planets and Moons of the Solar System in Art, Myth and History 历史神话艺术中的太阳系的太阳，行星，矮行星和卫星。Amazon, 2015. ISBN-10: 1515169480; ISBN-13: 978-1515169482

Volume 3—A Selection of mainly Larger Asteroids, their Moons and Four Comets in Art, Myth and History 历史神话艺术中的大的小行星及其卫星和四个彗星。Amazon, 2015. ISBN-10: 1515290840; ISBN-13: 978-1515290841

Volume 4—Tarot-Mahjong related to the 5 Elements, the 8 Trigrams and other traditional series 塔罗麻将牌与五行， 八卦及其它传统元素的关联。Amazon, 2015. ISBN-10: 1511461373; ISBN-13: 978-1511461375

Volume 5—Tarot-Mahjong Images with Chinese and Western Elements, Planets, Moons and Stars 塔罗麻将牌—包含中西方五行，行星，卫星及星辰之间的关联. Sydney, Amazon, 2015. ISBN-10: 1508778736; ISBN-13: 9781508778738

Volume 5— Astronomical Pencil Drawings with Chinese and Western Elements, Planets, Moons and Stars 天文学方面的铅笔画—包含中西方五行，行星，卫星及星辰之间的关联。Amazon, 2016. ISBN-13: 978-1537192697; ISBN-10: 1537192698 (This is an updated edition of Tarot-Mahjong Images with Chinese and Western Elements)

Pre-Columbian Discoveries of the New World by Asians, Africans and Europeans and some Ancient Native American Voyages to Europe, Africa and Asia 亚非欧洲人在哥伦布之前发现美洲新大陆，以及古代美国原住民的亚非欧航行之旅。 Amazon, 2015. ISBN-10: 1514202190; ISBN-13: 978-1514202197

Chinese Nationalism and Politics in Indonesia 1900-1965 : 1900-1965 在印度尼西亚的华侨民族主义和政治运动。Amazon, 2016. ISBN-13: 978-1536802689; ISBN-10: 1536802689

Christianity—a Restatement of Greco-Roman Beliefs 基督教—古希腊罗马信仰的另一种形式。Amazon, 2016. The Rise of the Mauryan Empire and India's Relations with the Ancient Greek World 孔雀王国的崛起以及印度和古希腊的关系。Amazon, 2016. Was Noah a Woman? 诺亚是不是女人? Amazon, 2016. (These 3 are all in the same cover) ISBN-13: 978-1536919851; ISBN-10: 1536919853

Chinese Lunar Calendar—Gregorian Calendar Conversion Tables for the years 1804/1805—2100 中国公历—农历换算表 1804/1805 —2100 年。 Amazon, 2016. ISBN-13: 978-1539320234; ISBN-10: 1539320235

The Invention of Playing Cards in China, their Arrival in Europe and their Role in the Development of Mahjong 中国纸牌的发明，以及在欧洲的传播，和对麻将发展的影响。Amazon, 2016. ISBN-13: 978-1539507024; ISBN-10: 1539507025.

My Life Story during the 20th Century by Gladys Muriel Goodman nee Wiggins — Autobiography of an Australian Woman: An autobiography with historical references, typed with minor editing by John Oxenham Goodman from the original manuscript presented to him by his mother Gladys Muriel Goodman nee Wiggins (29 October 1903-20 September 2002) ISBN- 13: 978-1537691558; ISBN- 10: 1537691554; Amazon, 2016.

New Concepts in Playing Cards and Tarot: Five Newly Designed Packs of Cards with Chinese Symbolism 新发明之融入中国元素的五副纸牌和塔罗牌。Amazon 2016. ISBN-13: 978-1541023574 ; ISBN-10: 1541023579.

Chinese Culture and History with Pictures and Descriptions—Everything from the Eight Immortals and Auspicious Animals to Military Badges, Dominoes and Chess 中国文化和历史的描述及图片—有关八仙，瑞兽，古代官员服饰，多米诺骨牌及象棋。Amazon 2017. ISBN-13: 978-1542881869; ISBN-10: 1542881862.

China's Borders with Russia in the Northeast: Unequal Treaties and Large Territorial Losses 中国东北部与俄罗斯的边界线：不公平条约和大量领土损失. Amazon 2017.ISBN-13: 978-1545358467; ISBN-10: 154535846X.

A Chinese Version of Tarot explained in Detail: Five Suits based on the Five Elements: 中文版本塔罗牌及详解 —中国五行所对应的塔罗牌的五个组成部分. Amazon 2017. ISBN-13: 978-1546469636; ISBN-10: 154646963X

Fortune Stars, Immortals and Dragon Sons: Three Newly Designed Packs of Cards based on Chinese Cultural Concepts: 五星，八仙和龙九子：根据中国文化发明的三副新纸牌. Amazon 2017. Amazon 2017. ISBN- 13: 978-1973812500; ISBN- 10: 1973812509.

Foreword

My interest in numerals from different writing systems goes back to the 1970s when I began to study Japanese, Chinese, Indonesian, Javanese and Arabic. Around this time, having produced numerous pencil drawings and oil paintings I began to appreciate the artistic merit of symbols and symbolic representations of historical events. This led to my attempts to design playing cards based on new concepts, radically different from the usual Hearts, Diamonds, Clubs and Spades. My new designs were at first based on the planets and moons of the solar system and the Greek and Roman mythology surrounding their names. Then aspects of Chinese history and culture came to play a major role in my designs. The Five Elements of Chinese Daoism represented five suits in a Chinese version of Tarot. The emphasis then shifted to the Eight Immortals and their Eight Ritual Implements; the Eight Auspicious Buddhist Treasures; the Eight Trigrams of Earlier Heaven and Later Heaven; the Eight Martial Arts Trigrams; the Nine Sons of the Dragon; the Nine Great Emperors representing the Nine Northern Dipper Stars and the nine Imperial Chinese civilian and military ranks. Some of these new card games were published by Amazon in 2016 under the title *New Concepts in Playing Cards and Tarot: Five Newly Designed Packs of Cards with Chinese Symbolism*. Then in 2017 I went on to publish *A Chinese Version of Tarot explained in Detail: Five Suits based on the Five Elements.* This was followed by *Fortune Stars, Immortals and Dragon Sons: Three Newly Designed Packs of Cards based on Chinese Cultural Concepts.*

Moving from packs of cards to numerals was an exciting path to follow. The elegantly curved shapes of many of the Asian numerals continued to fascinate me and so did the geometrical forms of numerals based on the Phoenician alphabet with their vertical, horizontal and diagonal lines. Symbols are employed on playing cards which are also numbered (with the exception of Japanese Flower Cards) and the digits of the world's numeral systems are themselves symbols which can be enlarged to playing-card size. The numerals in this book are therefore in a format suitable for imprinting on playing cards. As a student and teacher of Asian languages I learnt long ago that characters and numbers in a foreign script need to be greatly enlarged to be appreciated and remembered by those newly encountering them. On beginning my Chinese studies I spent many hours looking at characters through a magnifying glass and later enlarging them to a size-20 font on a computer to facilitate reading. Large Chinese characters on shop fronts are easily recognized but in a size-11 font in a pamphlet, the eyes are strained, particularly if one has less than perfect vision requiring the wearing of glasses.

The other aspect of this book is the history of symbols used to help human beings record numbers. These numbers are very important in almost every branch of learning. They are the basic essentials of trade and commerce as well as architecture, building and construction. Then there are the fields of mathematics and astronomy as well as almost every other branch of learning. The origins of numbers and mathematics in ancient America, China, Egypt, Mesopotamia, Greece and India give fascinating historical glimpses of mankind's progress to modernity.

John Oxenham Goodman, Shenyang City, Liaoning Province, China

A Brief Biography of the Author

I was born in Australia in 1941. As a young man I worked in clerical positions in government departments but found these jobs routine and unchallenging. I studied Spanish and German and then travelled extensively in Western Europe crossing into Soviet occupied Berlin and climbing a mountain in Norway to view the midnight sun. These and many other travel experiences opened my eyes to the wider world and gave me a broader perspective than that available in the geographical remoteness of Australia. My extensive travels in Europe helped remove that sense of isolation from European and Western culture felt by many Australians.

On returning to Australia I found employment in the Australian National University Library. There I came in contact with many Asian people who worked in or frequented the university's Oriental Library and I developed an interest in Chinese and Japanese civilizations. In 1969 I enrolled in the Asian Studies Faculty, first studying Indonesian and Japanese. Eventually, in early 1973, I undertook an intensive course in spoken Chinese at the University of Canberra and then studied Classical Chinese at the Australian National University. I finished my Indonesian and Asian Studies majors and studied Javanese and Arabic while completing a reading course in Dutch. I graduated with a Bachelor of Arts Honours degree and later received graduate diplomas in Education and Librarianship going on to major in Japanese language at the University of New South Wales. With a lifelong interest in art and photography I undertook a course in painting and drawing at the TAFE (Technical and Further Education) campus in the Sydney suburb of Campbelltown.

I taught Indonesian and Japanese languages to students from year 7 to year 12 in Australian secondary schools as well as French and German to students beginning year 7. I also taught Indonesian native speakers in year 11 and year 12 classes where politically sensitive literature which had been banned or restricted in Indonesia was part of the curriculum. This was the Education Department's way of introducing them to the freedoms enjoyed in Australia. Students beginning Japanese coloured in simple line drawings of the Japanese Flower Cards which I had produced. This was part of their introduction to Japanese culture.

In the mid 1980s I started a series of drawings of the mythical figures which had given their names to the constellations, planets, moons and major asteroids. I was then working in the University of Sydney Library and had access to useful information on astronomy, Greco-Roman history and art, a prerequisite for someone wanting to depict the legendary heroes after whom the heavenly bodies were mostly named. At first I wrote about each of these hundreds of mythical or historical figures and then began the drawings.

After retiring from the New South Wales Education Department I studied the Teaching of English as a Second Language at the Australian Catholic University in Sydney. I went on to teach English to newly arrived adult migrants at a college in the Sydney suburb of Hurstville. At that time Hurstville was Australia's largest Chinatown and of course my students came mainly from China. This further enhanced my knowledge and appreciation of Chinese culture.

While living in China I found displays in China's provincial museums to be a wonderful source of information on the ancient past. I travelled by train to most of China's provinces paying special attention to Buddhist, Daoist and Confucian temples. There I saw statues, engravings, paintings and art work which, when added to the enlightenment gained from museum visits, gave me fresh understanding of this 5000 year old civilization.

John Oxenham Goodman, Shenyang City, Liaoning Province, China

Contents

Circled Roman or Greek letters have been placed at the upper left of each of these 48 sets of numerals as a means of distinguishing them from one another.

Greek letters symbolizing the numerals: ϟ koppa; Λ lambda; Θ theta; β beta; α alpha; ξ xi; ϡ sampi; Ω omega; Δ delta; Γ gamma; Π pi; Φ phi; Ψ psi; Σ sigma.

South American Knot Numerals 南美洲结绳数字

There is a possibility that the civilization of ancient China influenced the invention of numerals in South America. We know that in Peru the use of *quipu* knots of various shapes, sizes and colours were used to encode information. These knots could represent numerals one to ten as well as the hundreds and thousands and were spaced apart and arranged in a base 10 positional system. In ancient China around 2650 BC we are told that when the Yellow Emperor 黄帝 Huáng Dì found that tying knots to record information was unsatisfactory, he ordered his official historian Cāng Jié 仓颉 to create characters for writing. Knotted cords were used in China more than 4,000 years ago, even before the time of the legendary emperor Fú Xī 伏羲, and continued in use until the end of the Qing Dynasty in 1911. They were also used in ancient Hawaii.

1

Quipu Knot Numerals of South American Quechua

南美洲的盖丘亚奇普文字(结绳文字)

Huk

2

Quipu Knot Numerals of South American Quechua

南美洲的盖丘亚奇普文字(结绳文字)

Iskay

3

Quipu Knot Numerals of South American Quechua

南美洲的盖丘亚奇普文字(结绳文字)

Kinsa

4

Quipu Knot Numerals of South American Quechua

南美洲的盖丘亚奇普文字(结绳文字)

Tawa

5 Quipu Knot Numerals of South American Quechua

南美洲的盖丘亚奇普文字(结绳文字)

Pisqa

6 Quipu Knot Numerals of South American Quechua

南美洲的盖丘亚奇普文字(结绳文字)

Suqta

7 Quipu Knot Numerals of South American Quechua

南美洲的盖丘亚奇普文字(结绳文字)

Qanchis

8 Quipu Knot Numerals of South American Quechua

南美洲的盖丘亚奇普文字(结绳文字)

Pusaq

9

Quipu Knot Numerals of South American Quechua

南美洲的盖丘亚奇普文字(结绳文字)

Isqun

10

Quipu Knot Numerals of South American Quechua

南美洲的盖丘亚奇普文字(结绳文字)

Chunka

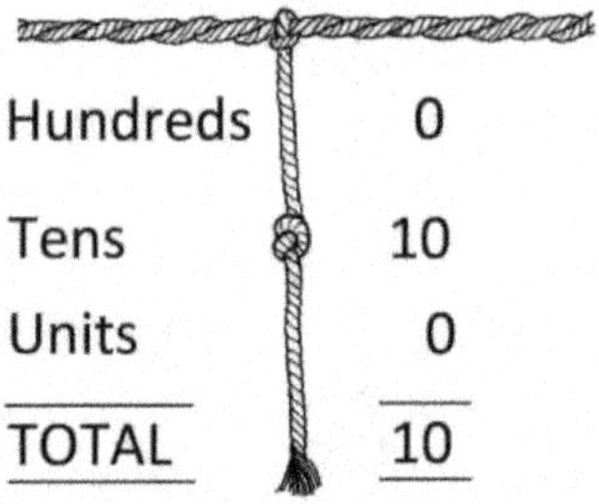

11

Quipu Knot Numerals of South American Quechua

南美洲的盖丘亚奇普文字(结绳文字)

Chunka hukniyuq

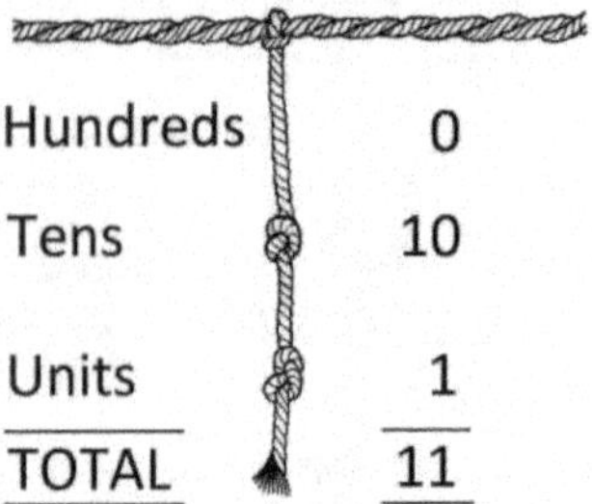

12

Quipu Knot Numerals of South American Quechua

南美洲的盖丘亚奇普文字(结绳文字)

Chunka iskayniyuq

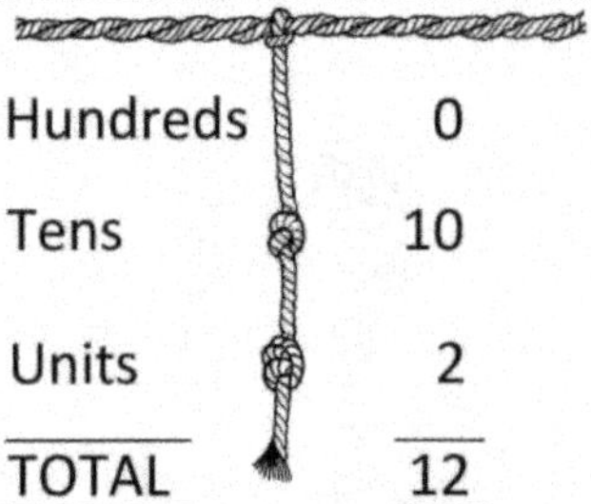

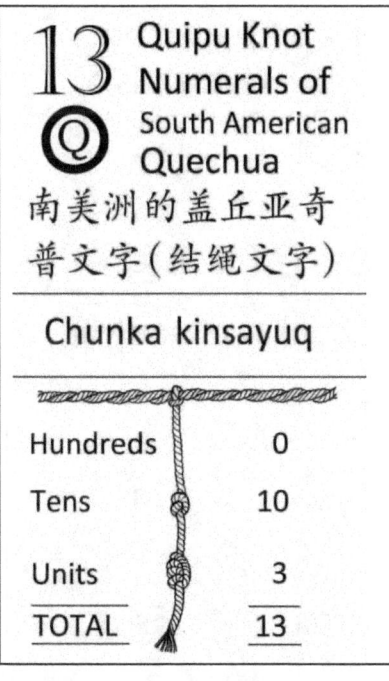

13 Quipu Knot Numerals of South American Quechua

南美洲的盖丘亚奇普文字(结绳文字)

Chunka kinsayuq

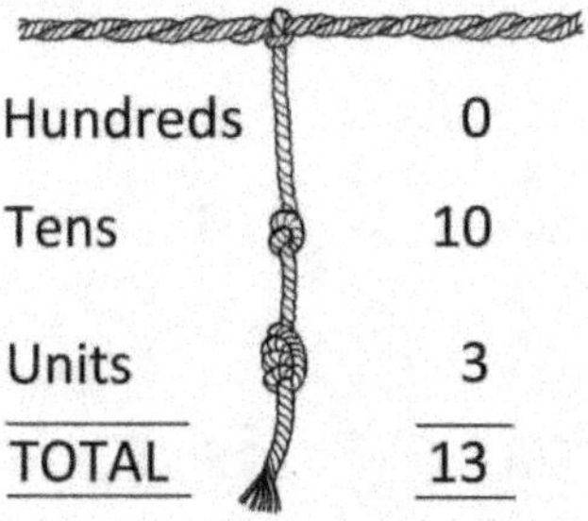

Hundreds	0
Tens	10
Units	3
TOTAL	13

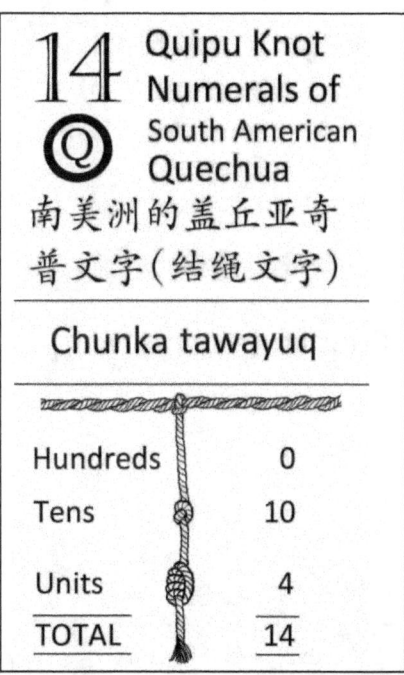

14 Quipu Knot Numerals of South American Quechua

南美洲的盖丘亚奇普文字(结绳文字)

Chunka tawayuq

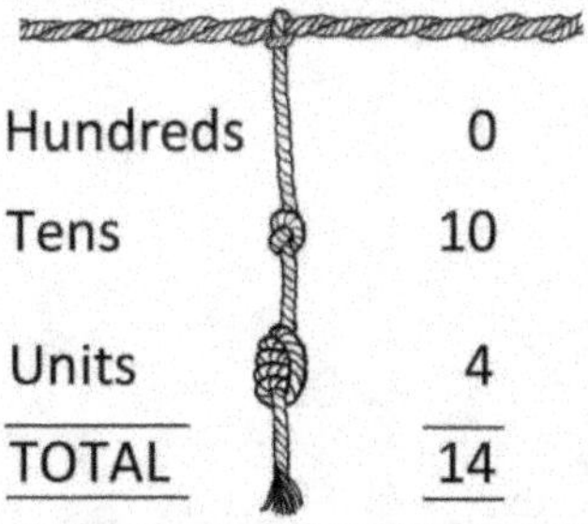

Hundreds	0
Tens	10
Units	4
TOTAL	14

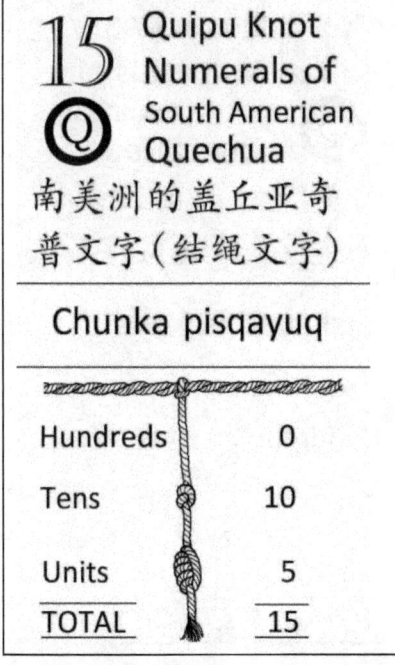

15 Quipu Knot Numerals of South American Quechua

南美洲的盖丘亚奇普文字(结绳文字)

Chunka pisqayuq

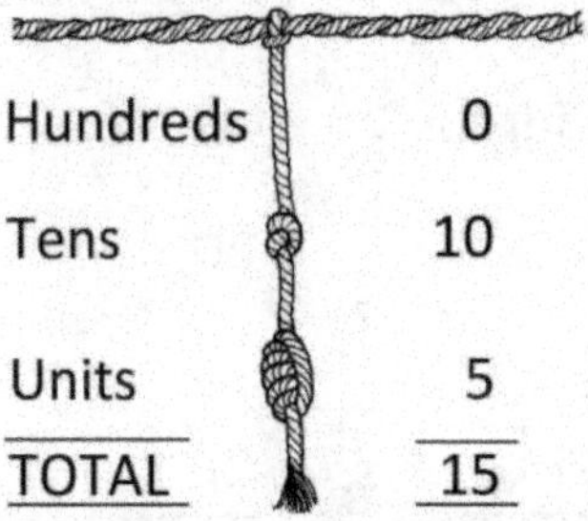

Hundreds	0
Tens	10
Units	5
TOTAL	15

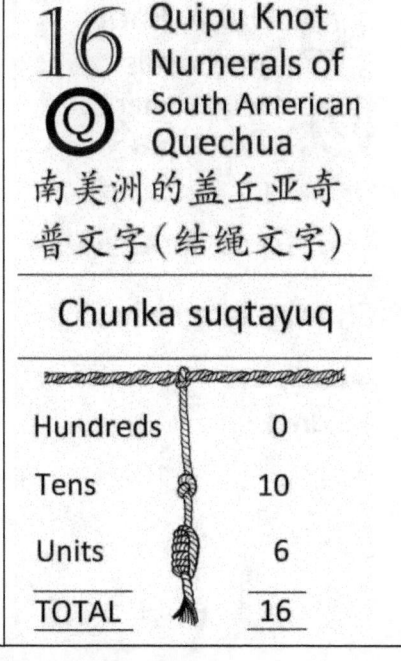

16 Quipu Knot Numerals of South American Quechua

南美洲的盖丘亚奇普文字(结绳文字)

Chunka suqtayuq

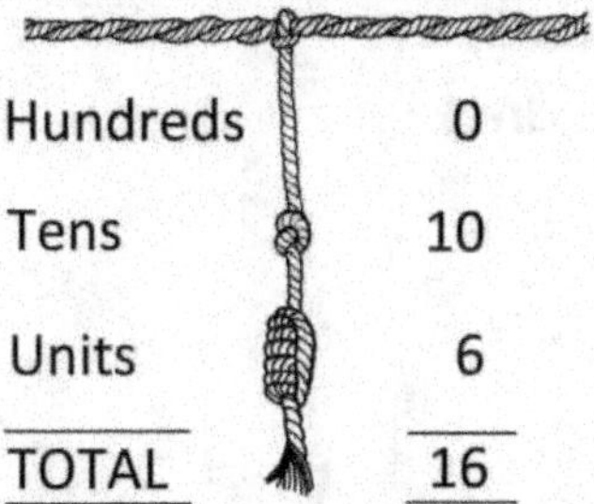

Hundreds	0
Tens	10
Units	6
TOTAL	16

17 Quipu Knot Numerals of South American Quechua

南美洲的盖丘亚奇普文字(结绳文字)

Chunka qanchisniyuq

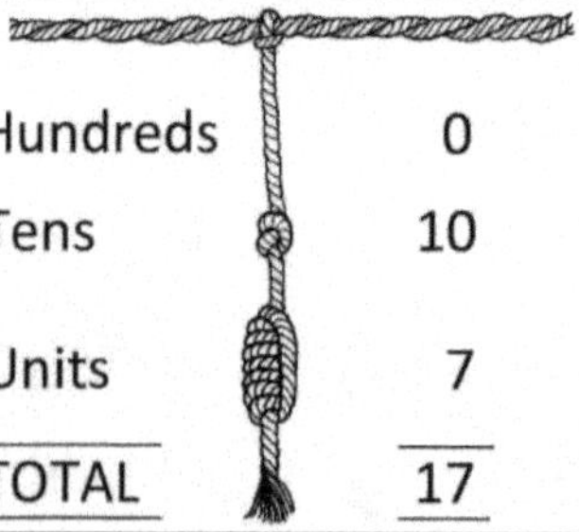

Hundreds	0
Tens	10
Units	7
TOTAL	17

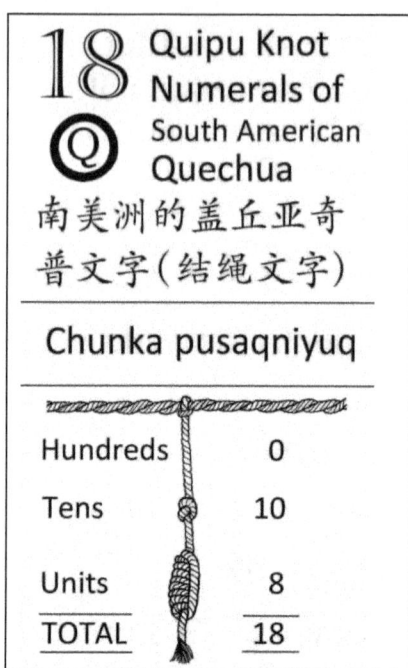

18 Quipu Knot Numerals of South American Quechua

南美洲的盖丘亚奇普文字(结绳文字)

Chunka pusaqniyuq

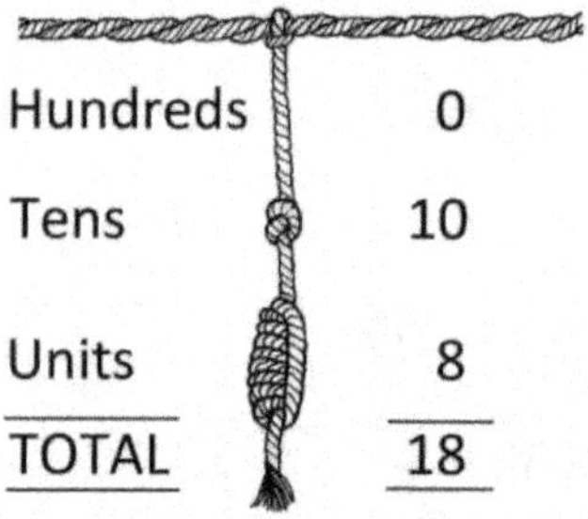

Hundreds	0
Tens	10
Units	8
TOTAL	18

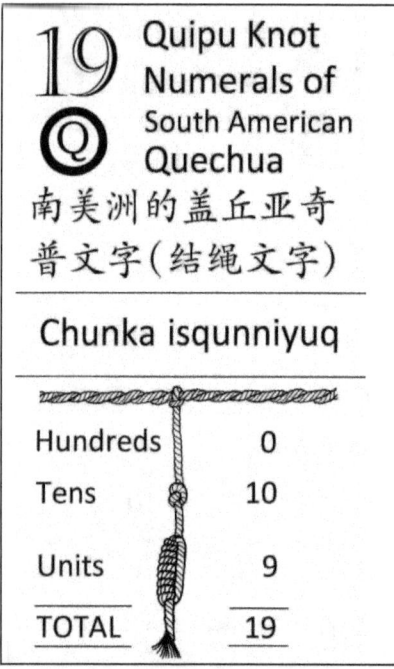

19 Quipu Knot Numerals of South American Quechua

南美洲的盖丘亚奇普文字(结绳文字)

Chunka isqunniyuq

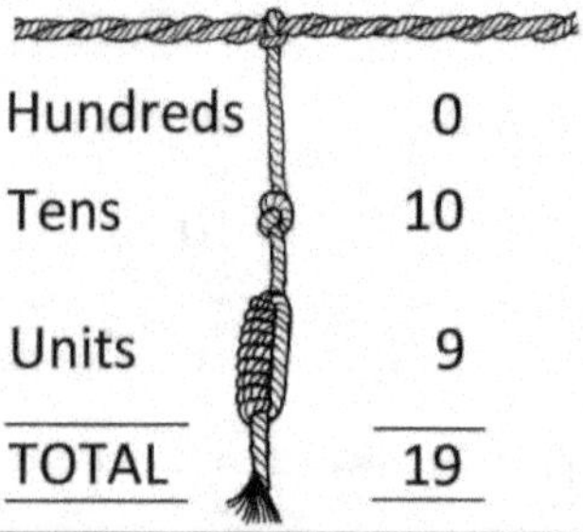

Hundreds	0
Tens	10
Units	9
TOTAL	19

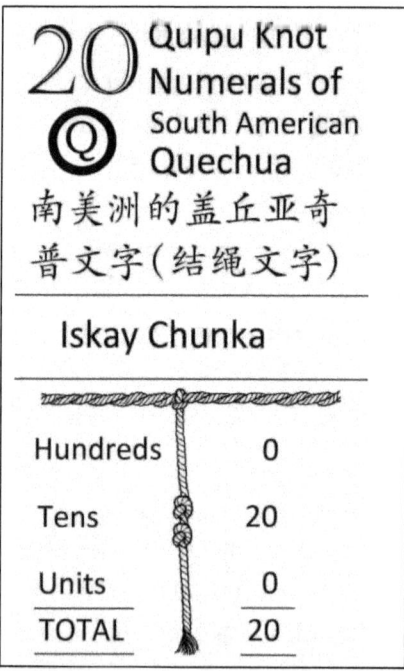

20 Quipu Knot Numerals of South American Quechua

南美洲的盖丘亚奇普文字(结绳文字)

Iskay Chunka

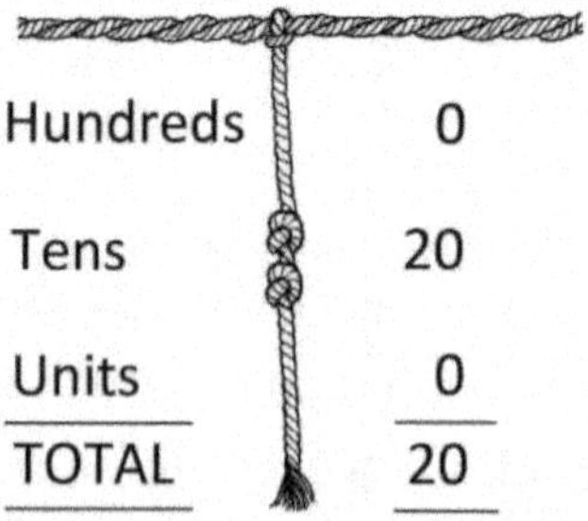

Hundreds	0
Tens	20
Units	0
TOTAL	20

50 Quipu Knot Numerals of South American Quechua

南美洲的盖丘亚奇普文字(结绳文字)

Pisqa chunka

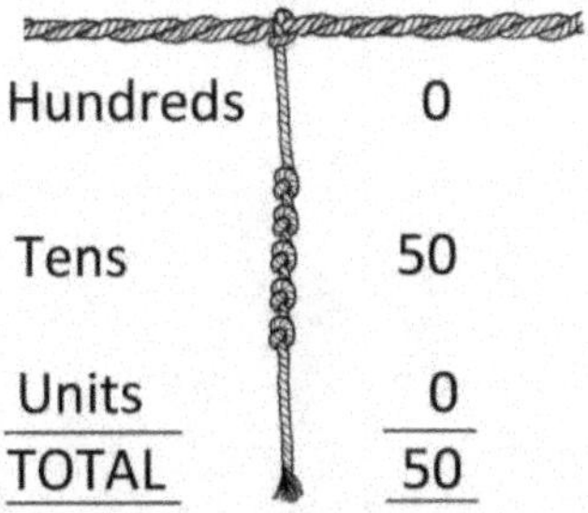

Hundreds	0
Tens	50
Units	0
TOTAL	50

100 Quipu Knot Numerals of South American Quechua

南美洲的盖丘亚奇普文字(结绳文字)

Pachak

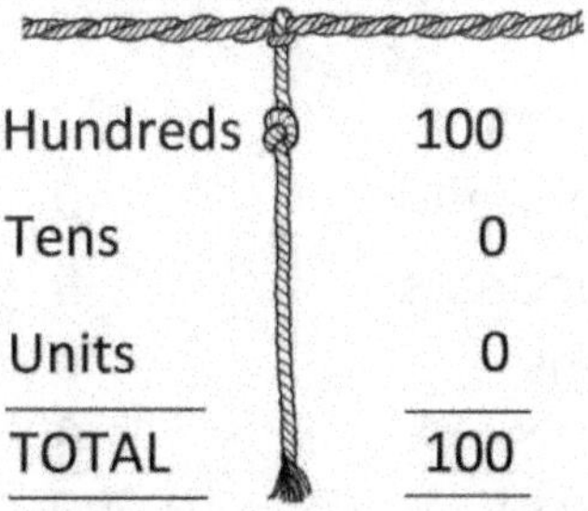

Hundreds	100
Tens	0
Units	0
TOTAL	100

500 Quipu Knot Numerals of South American Quechua

南美洲的盖丘亚奇普文字(结绳文字)

Pisqa Pachak

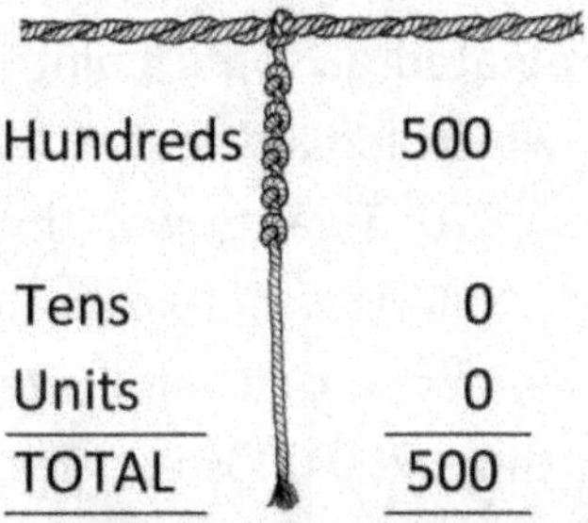

Hundreds	500
Tens	0
Units	0
TOTAL	500

1,000 Quipu Knot Numerals of South American Quechua

南美洲的盖丘亚奇普文字(结绳文字)

Waranqa

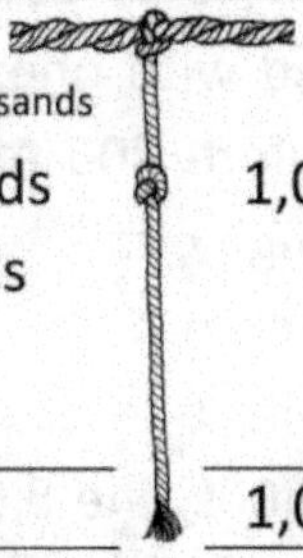

Tens of Thousands	0
Thousands	1,000
Hundreds	0
Tens	0
Units	0
TOTAL	1,000

10,000 Quipu Knot Numerals of South American Quechua

南美洲的盖丘亚奇普文字(结绳文字)

Chunka waranqa

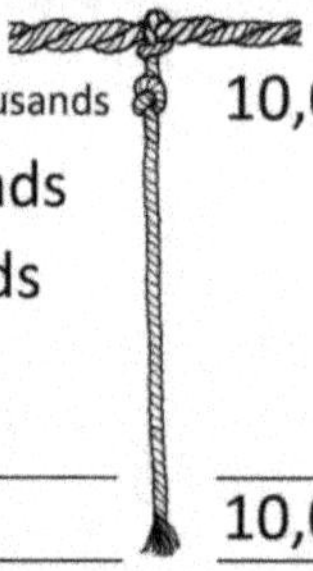

Tens of Thousands	10,000
Thousands	0
Hundreds	0
Tens	0
Units	0
TOTAL	10,000

Mayan Numerals 玛雅数字

The Mayans developed a base-twenty number system about 300 BC. Numbers one to four were represented by a corresponding number of dots and five was a horizontal bar. The dots and bar apparently symbolized beans and a bean pod. Six was represented by a dot above a bar, seven by two dots above a bar and so on up to nine. Ten was represented by two bars with dots placed above them for numbers 11 to 14. Fifteen was three bars with dots above for numbers 16 to 19. Twenty was again a single dot which could be confused with one. This problem was eliminated at some point in time before 400 AD when a conch shell, which looked something like a rugby ball, was used as a symbol for zero. Twenty was then represented by a dot above a conch shell. Numbers ran vertically with units 1-19 at the bottom. The 20s were above them, followed by the 400s, the 8,000s, the 160,000s and the 3,200,000s etc. Thus the Mayans had developed a base 20 positional numbering system superior to that of the Babylonians whose final zero was a blank space. The Mayan numbers 1 to 19 could also be symbolized by face glyphs for ceremonial purposes. A remaining deficiency was that numbers less than 20 were represented by combinations of bars and dots rather than single glyphs. The Babylonian system also suffered from the same problem which reduced the usefulness of both systems for mathematical calculations.

Although the numerals of the Maya are base 20, they are arranged in groups of five. We also find this grouping of 5 numerals in the Khmer language of Cambodia and some scholars believe that Khmer numerals were originally base 20. When we compare the 10th century seven-tier step pyramid of Koh Ker in northern Cambodia (120 kilometres from Angkor Wat) with similar structures in Mexico and Guatemala, the possibility of cultural transmission (to or from Mexico) arises.

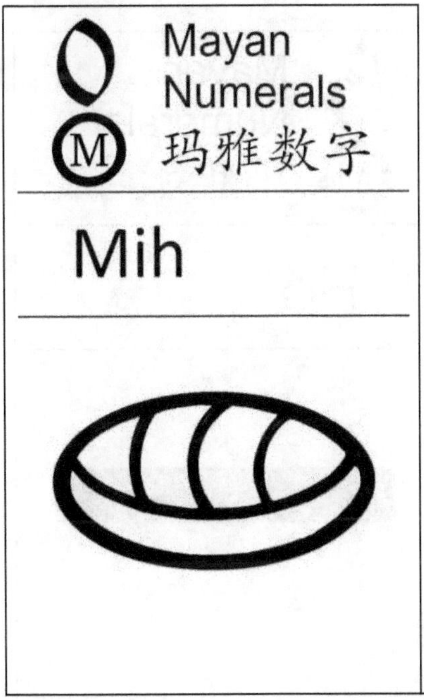

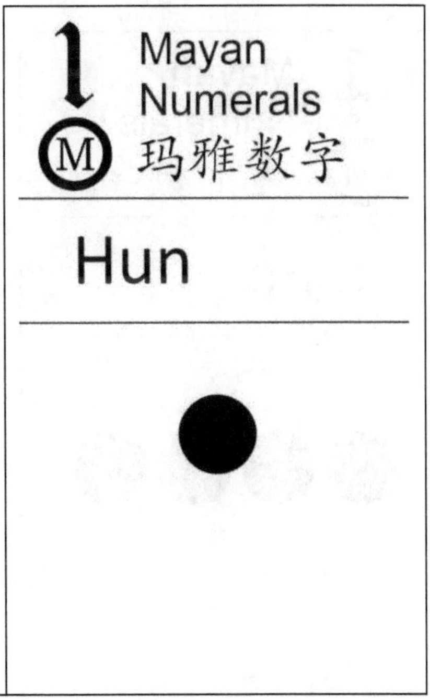

2 Mayan Numerals
M 玛雅数字

Ka'ah

3 Mayan Numerals
M 玛雅数字

Óox

4 Mayan Numerals
M 玛雅数字

Kan

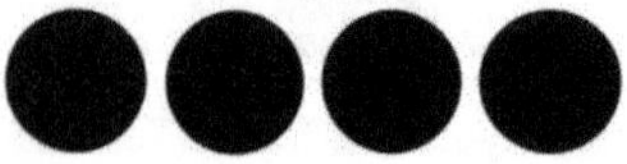

5 Mayan Numerals
M 玛雅数字

Ho'

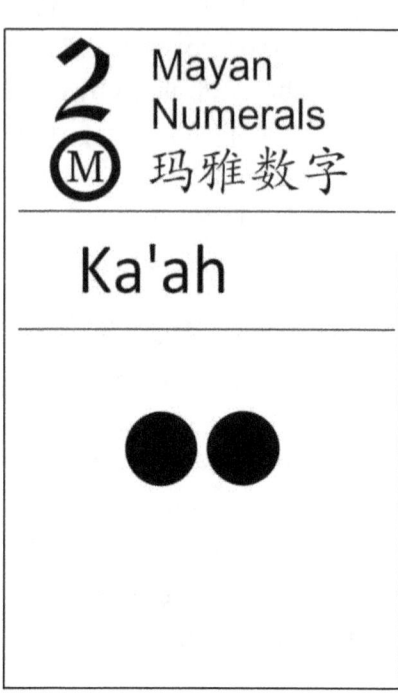

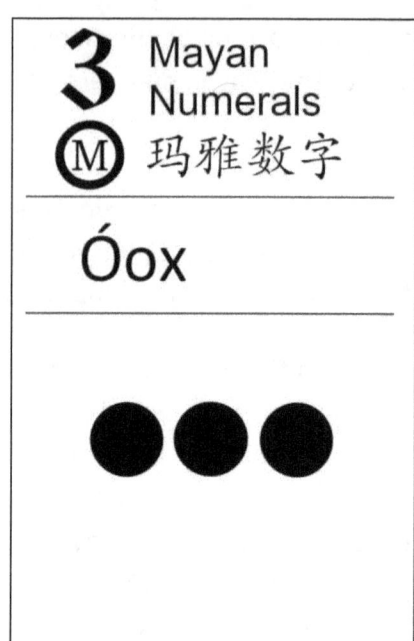

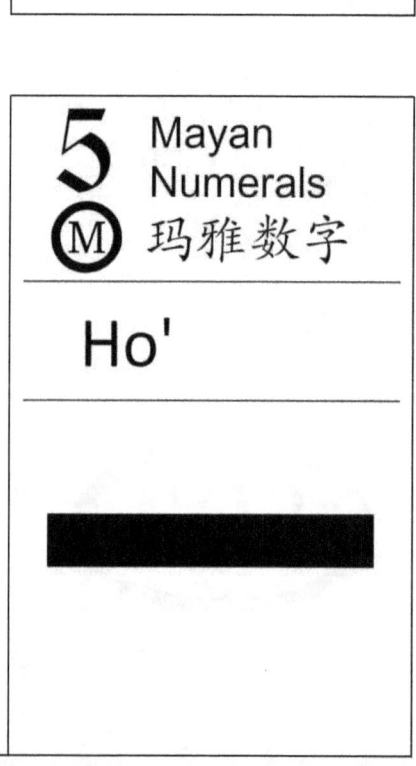

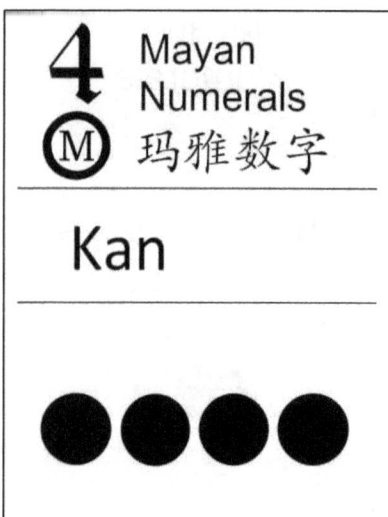

6 Mayan Numerals
Ⓜ 玛雅数字

Wak

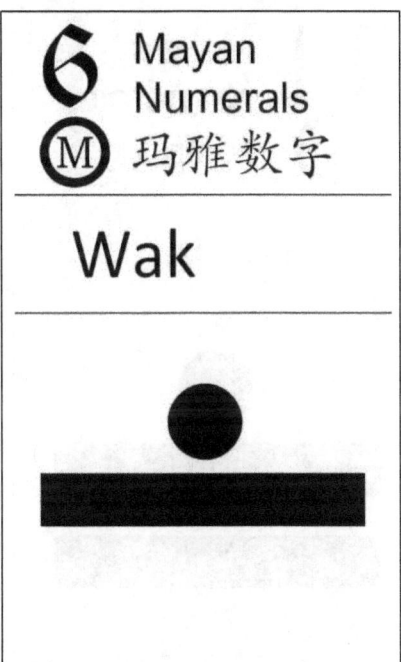

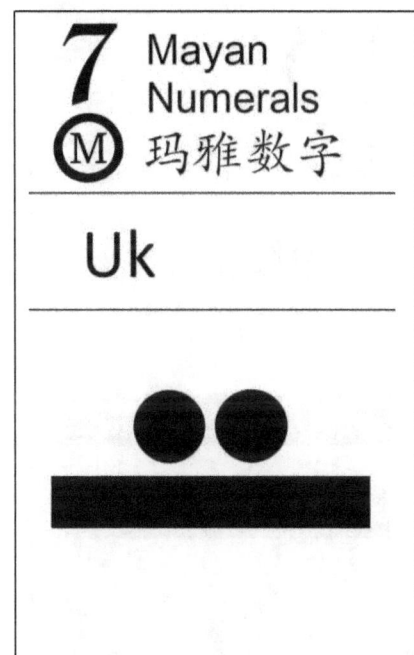

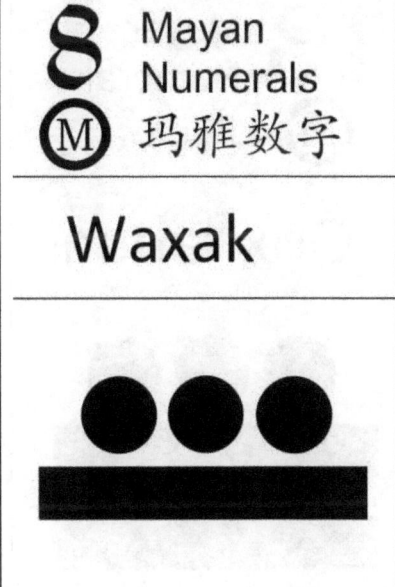

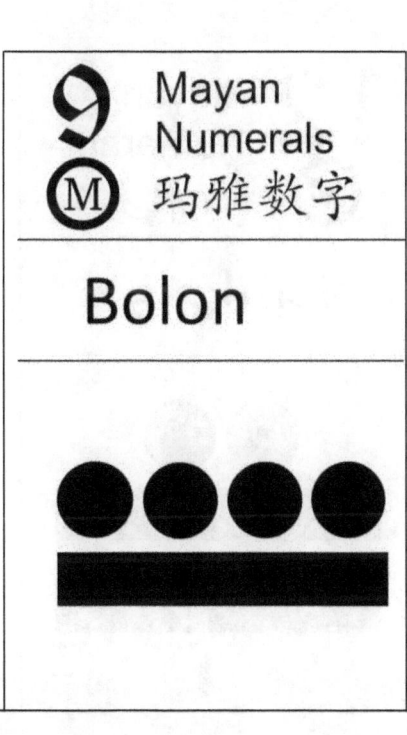

7 Mayan Numerals
Ⓜ 玛雅数字

Uk

8 Mayan Numerals
Ⓜ 玛雅数字

Waxak

9 Mayan Numerals
Ⓜ 玛雅数字

Bolon

10 Mayan Numerals
Ⓜ 玛雅数字

Lahun

11 Mayan Numerals
Ⓜ 玛雅数字

Buluk

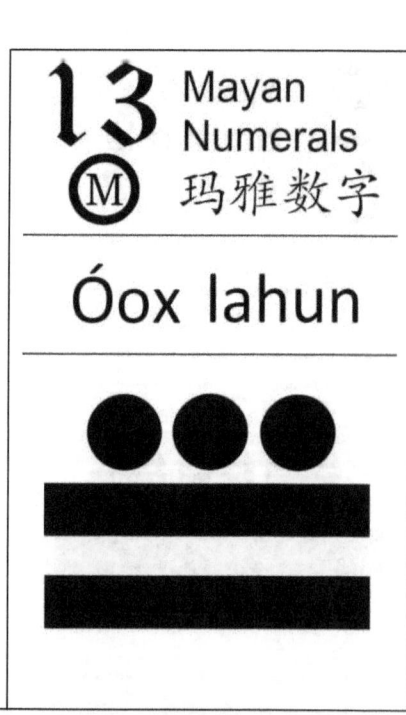

12 Mayan Numerals
Ⓜ 玛雅数字

Lahka'a

13 Mayan Numerals
Ⓜ 玛雅数字

Óox lahun

14 Mayan Numerals

Ⓜ 玛雅数字

Kan lahun

15 Mayan Numerals

Ⓜ 玛雅数字

Ho' lahun

16 Mayan Numerals

Ⓜ 玛雅数字

Wak lahun

17 Mayan Numerals

Ⓜ 玛雅数字

Uk lahun

18 Ⓜ Mayan Numerals 玛雅数字

Waxak lahun

19 Ⓜ Mayan Numerals 玛雅数字

Bolon lahun

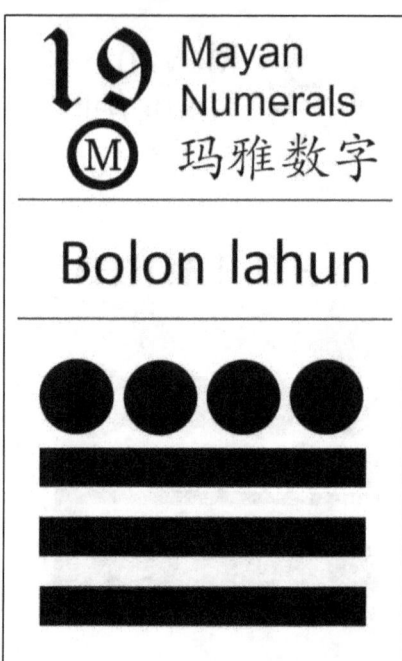

20 Ⓜ Mayan Numerals 玛雅数字

Hun k'áal

1 X 20 = 20

0 = 0

TOTAL = 20

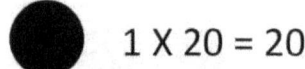

50 Mayan Numerals 玛雅数字

Lahun katak ka' k'áal

2 X 20 = 40

10 = 10

TOTAL = 50

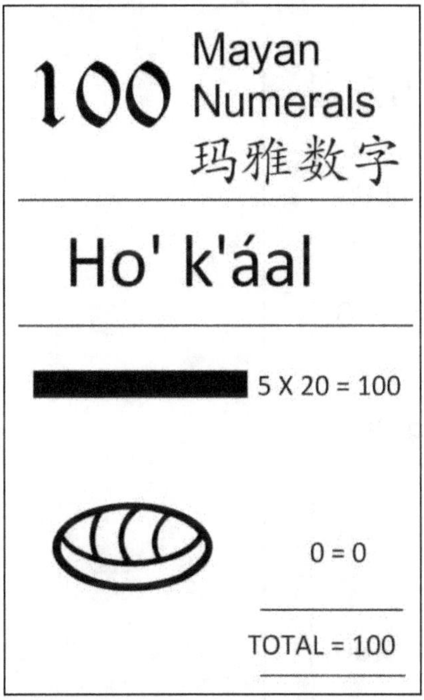
100
Mayan
Numerals
玛雅数字
Ho' k'áal
5 X 20 = 100
0 = 0
TOTAL = 100

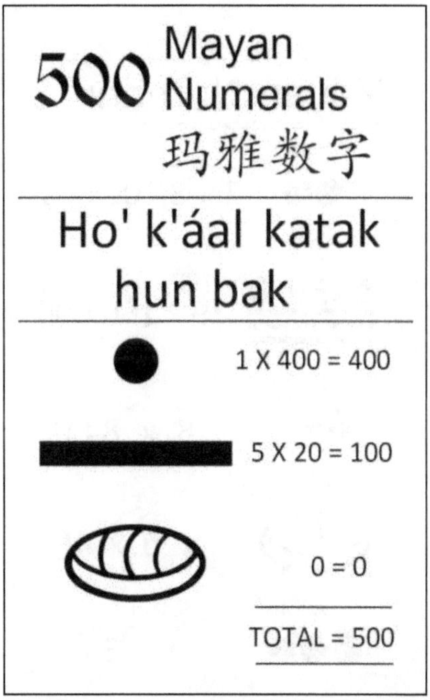
500
Mayan
Numerals
玛雅数字
Ho' k'áal katak
hun bak
1 X 400 = 400
5 X 20 = 100
0 = 0
TOTAL = 500

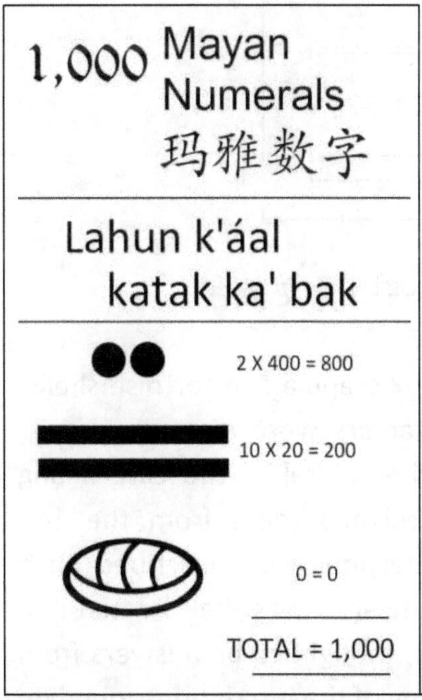
1,000
Mayan
Numerals
玛雅数字
Lahun k'áal
katak ka' bak
2 X 400 = 800
10 X 20 = 200
0 = 0
TOTAL = 1,000

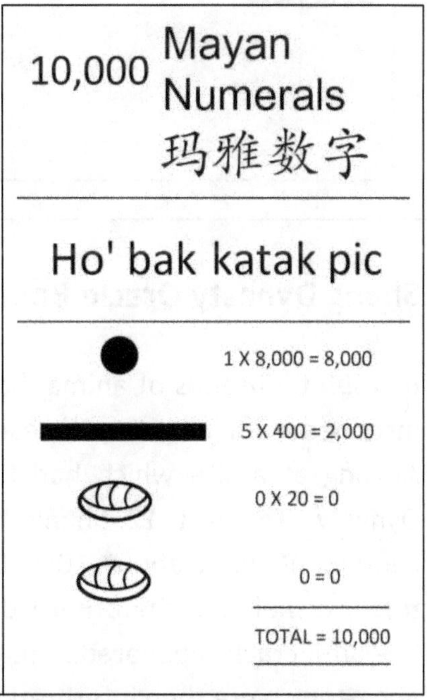
10,000
Mayan
Numerals
玛雅数字
Ho' bak katak pic
1 X 8,000 = 8,000
5 X 400 = 2,000
0 X 20 = 0
0 = 0
TOTAL = 10,000

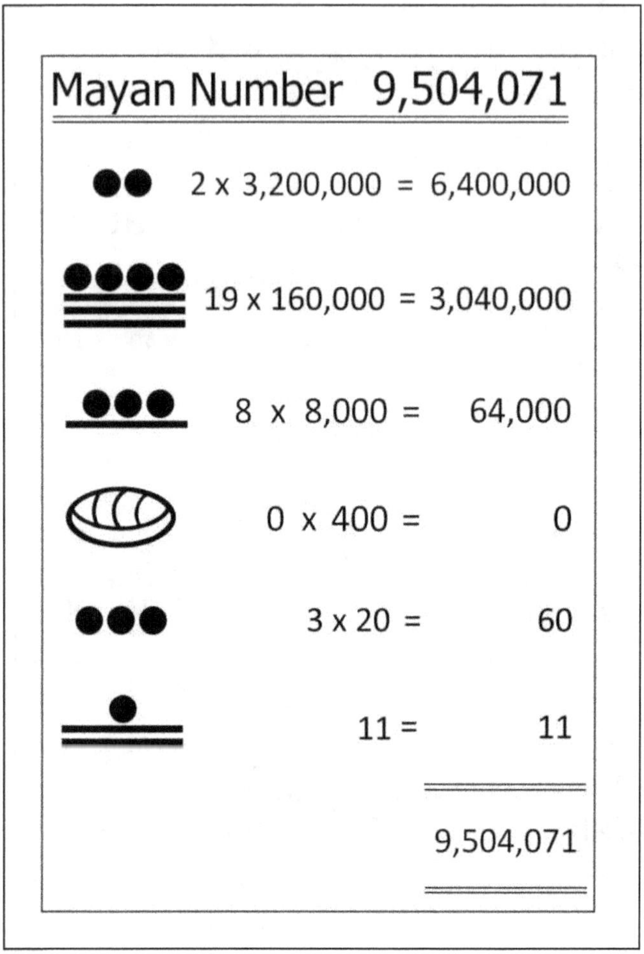

Shang Dynasty Oracle Bones 商朝甲骨文数字

In 1899 thousands of animal bones (ox scapula and tortoise shells) inscribed with ancient Chinese characters were discovered near Anyang at a site which had been the capital of the Late Shang Dynasty. The last 12 Shang kings reigned there from the 14th century BC until about 1045 BC. Questions were inscribed on a tortoise shell and the other side of the shell was then heated in a fire until cracks appeared. The cracks, thought to be answers from ancestors, were then interpreted. Many of the inscriptions found on the oracle bones were numerical in nature, giving such information as the number of soldiers killed or captured in battle, the number of sacrifices performed or the number of animals killed in hunting. This was a fully developed base 10 numerical system which was both multiplicative and additive but it was not a positional system and did not need a zero. The largest number found on oracle bones was 30,000 and there are several slightly different variants of some numerals.

The inscribed oracle bones are from the 14th Century BC but such a writing system must have evolved over hundreds of years and earlier examples of Chinese writing may yet be discovered by archaeologists. The story about the Yellow Emperor 黄帝 Huáng Dì tying knots to record information and finding this unsatisfactory, and then around 2650 BC ordering his historian Cāng Jié 仓颉 to create written characters, is an indication that Chinese writing and numerals long predate the oracle bones.

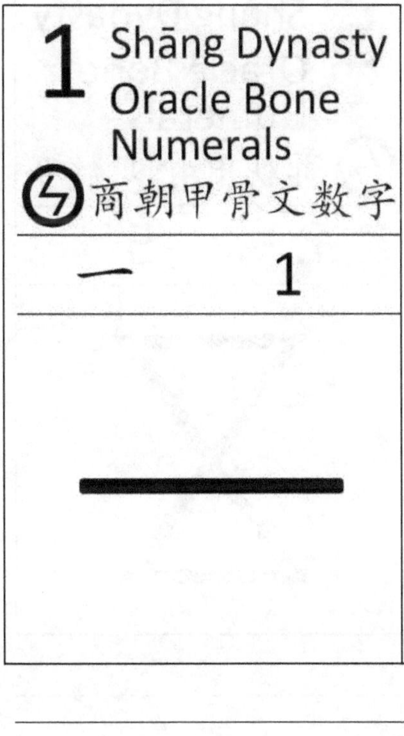

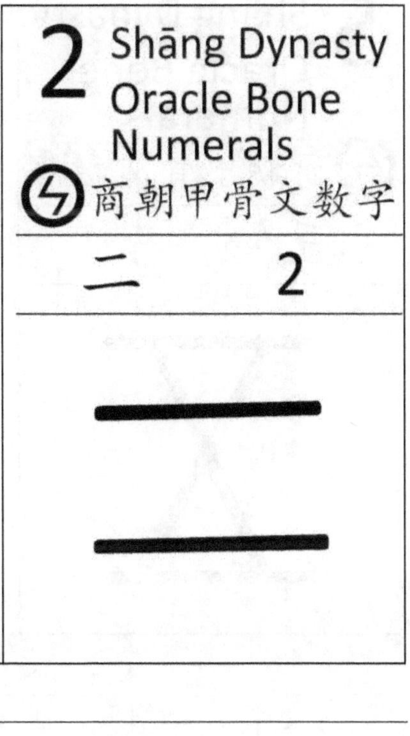

3 Shāng Dynasty Oracle Bone Numerals

商朝甲骨文数字

三 3

4 Shāng Dynasty Oracle Bone Numerals

商朝甲骨文数字

四 4

5 Shāng Dynasty Oracle Bone Numerals

商朝甲骨文数字

五 5

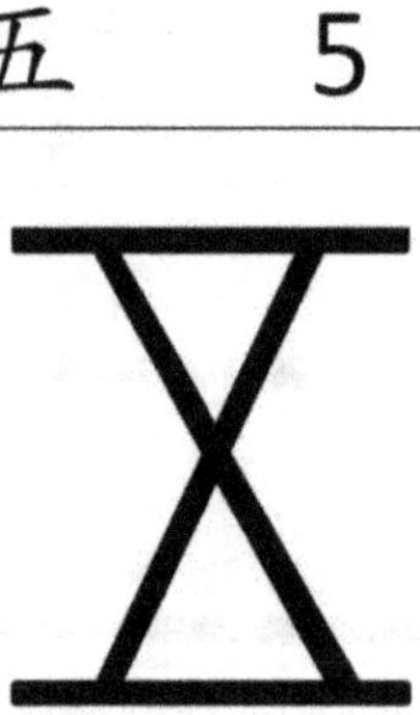

5 Shāng Dynasty Oracle Bone Numerals

商朝甲骨文数字

五 5

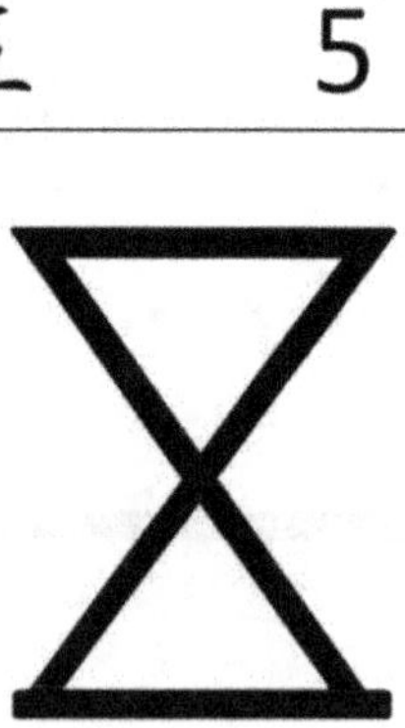

6 Shāng Dynasty Oracle Bone Numerals

商朝甲骨文数字

六 6

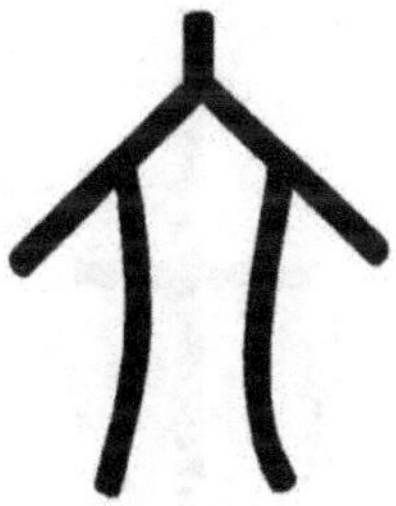

6 Shāng Dynasty Oracle Bone Numerals

商朝甲骨文数字

六 6

6 Shāng Dynasty Oracle Bone Numerals

商朝甲骨文数字

六 6

6 Shāng Dynasty Oracle Bone Numerals

商朝甲骨文数字

六 6

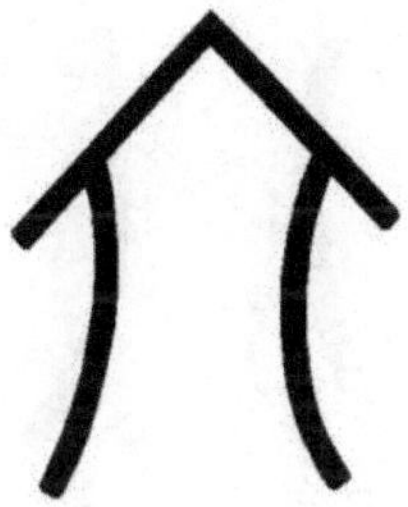

7 Shāng Dynasty Oracle Bone Numerals

商朝甲骨文数字

七 7

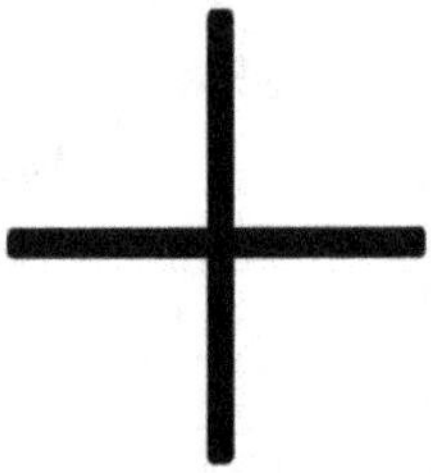

7 Shāng Dynasty Oracle Bone Numerals

商朝甲骨文数字

七 7

8 Shāng Dynasty Oracle Bone Numerals

商朝甲骨文数字

八 8

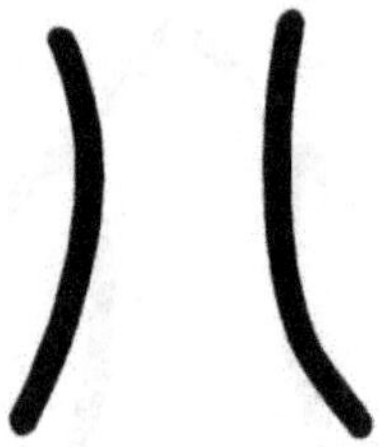

8 Shāng Dynasty Oracle Bone Numerals

商朝甲骨文数字

八 8

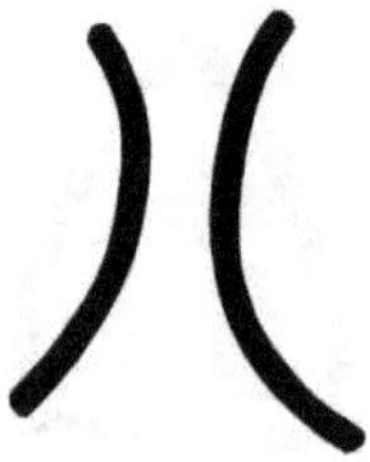

9 Shāng Dynasty Oracle Bone Numerals

商朝甲骨文数字

九 9

9 Shāng Dynasty Oracle Bone Numerals

商朝甲骨文数字

九 9

9 Shāng Dynasty Oracle Bone Numerals

商朝甲骨文数字

九 9

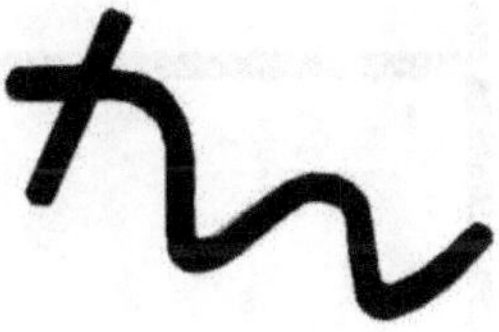

9 Shāng Dynasty Oracle Bone Numerals

商朝甲骨文数字

九 9

9 Shāng Dynasty Bronze Script Numeral
商周金文数字

九 9

10 Shāng Dynasty Oracle Bone Numerals
商朝甲骨文数字

十 10

11 Shāng Dynasty Oracle Bone Numerals
商朝甲骨文数字

十一 11

12 Shāng Dynasty Oracle Bone Numerals
商朝甲骨文数字

十二 12

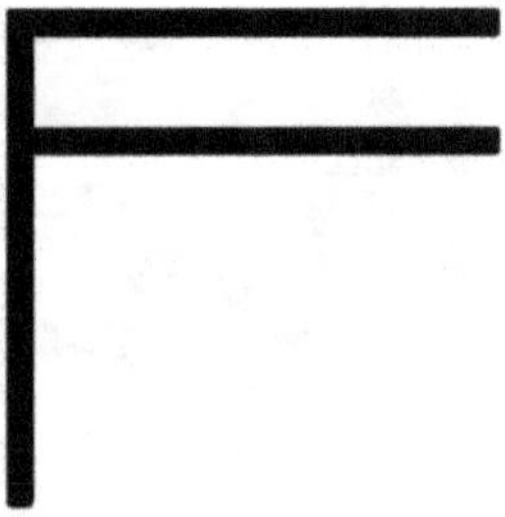

13 Shāng Dynasty Oracle Bone Numerals
商朝甲骨文数字

十三 13

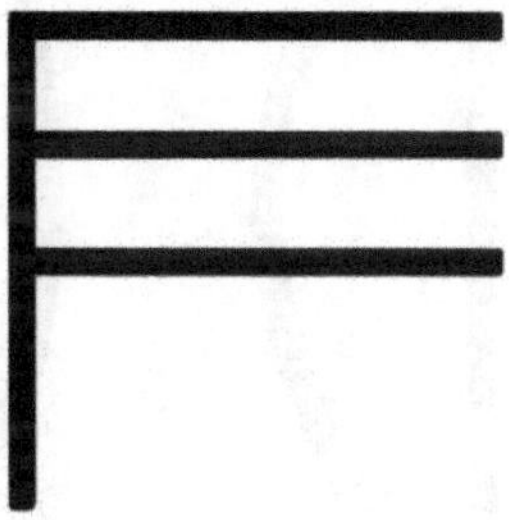

14 Shāng Dynasty Oracle Bone Numerals
商朝甲骨文数字

十四 14

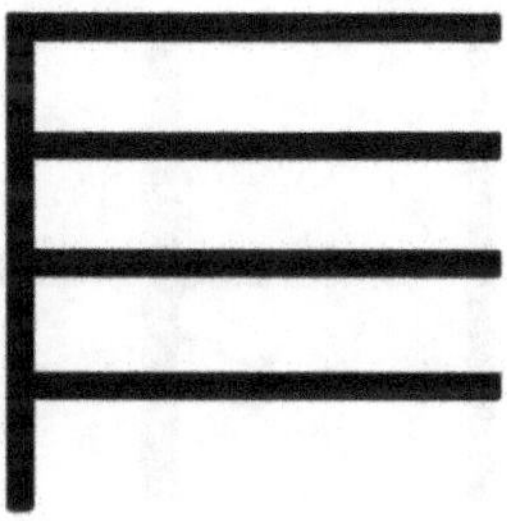

15 Shāng Dynasty Oracle Bone Numerals
商朝甲骨文数字

十五 15

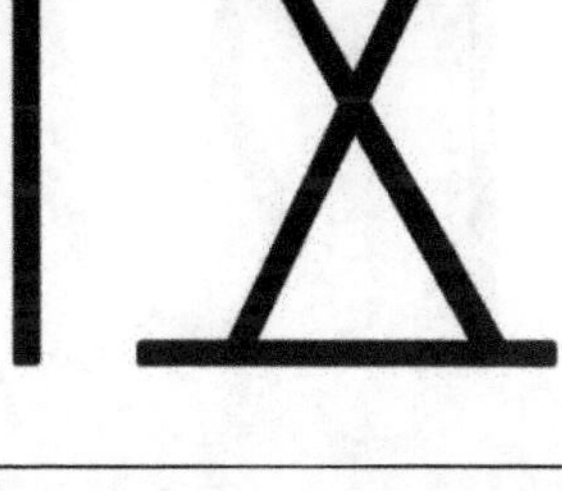

16 Shāng Dynasty Oracle Bone Numerals
商朝甲骨文数字

十六 16

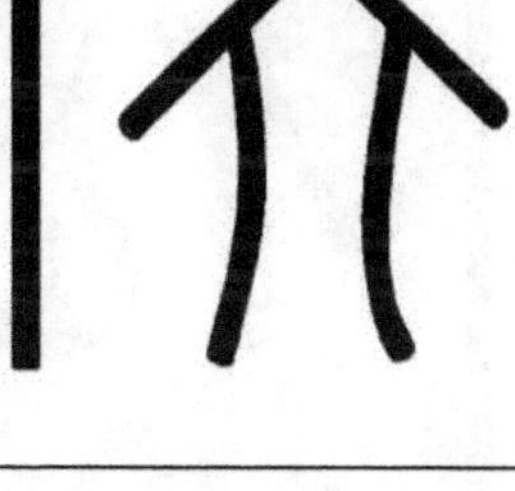

17 Shāng Dynasty Oracle Bone Numerals

商朝甲骨文数字

十七 17

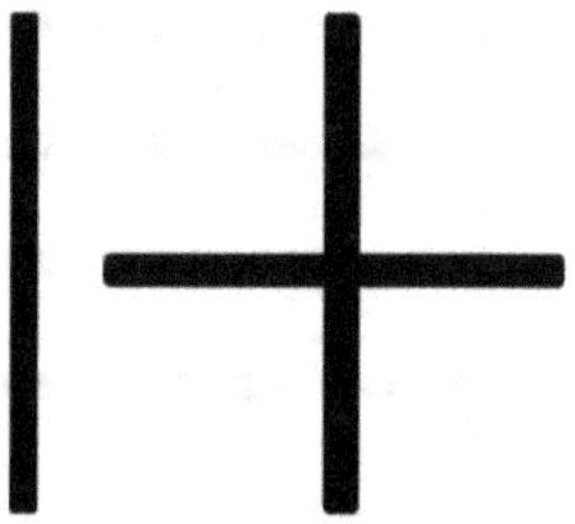

18 Shāng Dynasty Oracle Bone Numerals

商朝甲骨文数字

十八 18

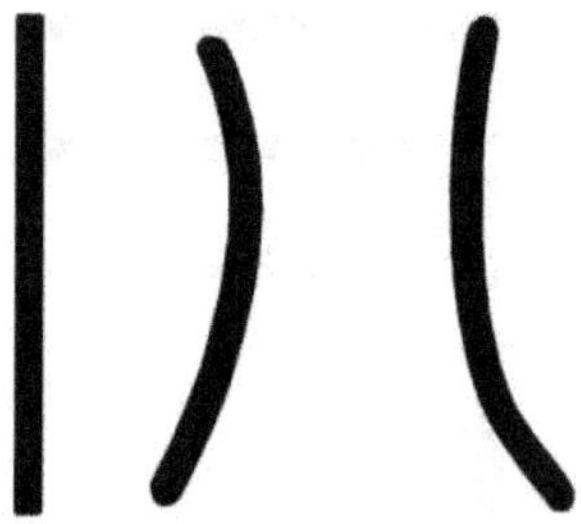

19 Shāng Dynasty Oracle Bone Numerals

商朝甲骨文数字

十九 19

19 Shāng Dynasty Oracle Bone Numerals

商朝甲骨文数字

十九 19

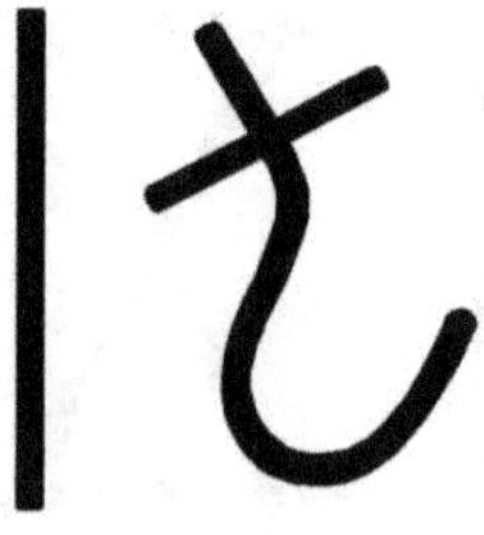

20 Shāng Dynasty Oracle Bone Numerals
商朝甲骨文数字

二十 20

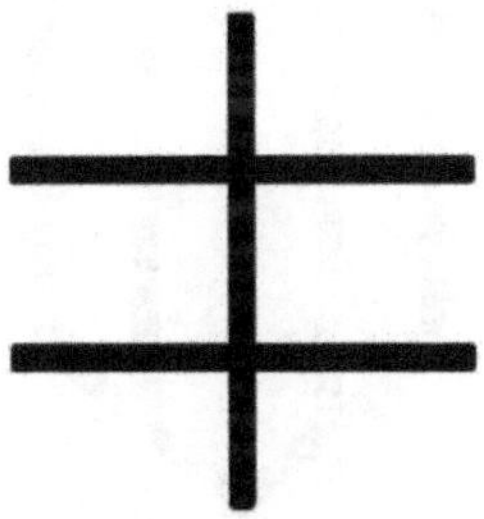

20 Shāng Dynasty Oracle Bone Numerals
商朝甲骨文数字

二十 20

30 Shāng Dynasty Oracle Bone Numerals
商朝甲骨文数字

三十 30

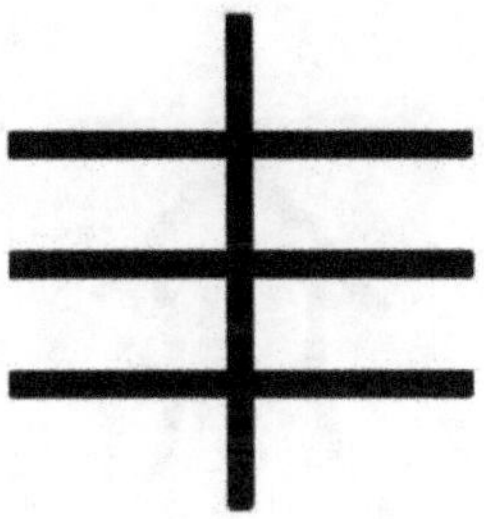

30 Shāng Dynasty Oracle Bone Numerals
商朝甲骨文数字

三十 30

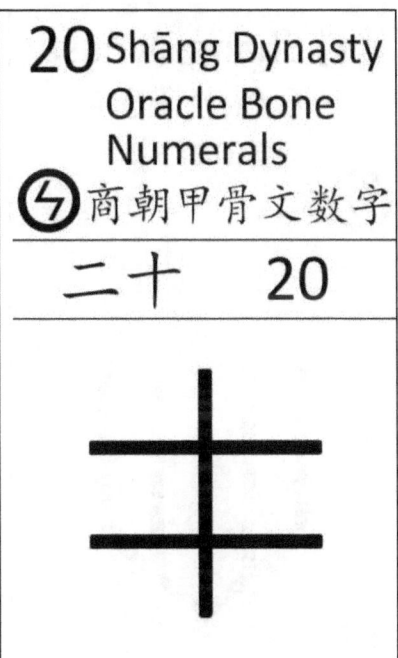

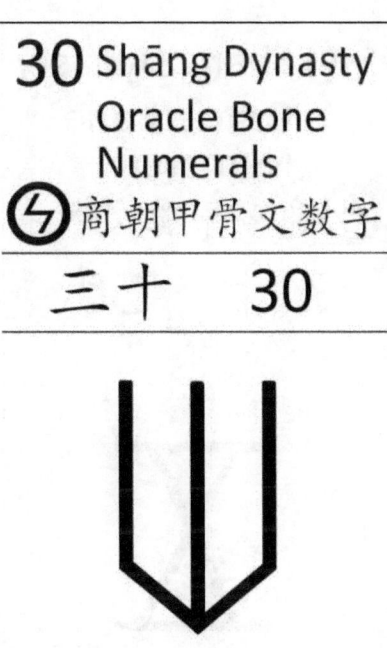

40 Shāng Dynasty Oracle Bone Numerals
商朝甲骨文数字

四十 40

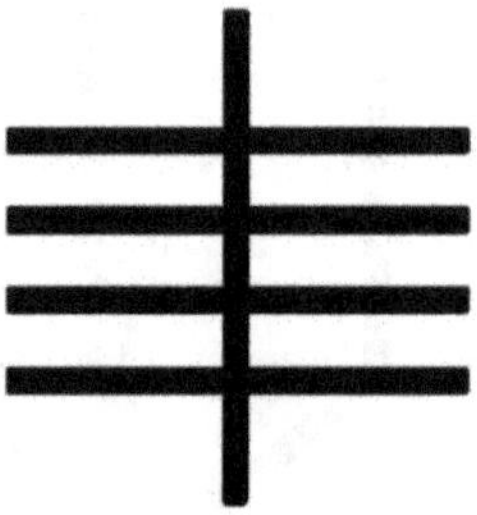

40 Shāng Dynasty Oracle Bone Numerals
商朝甲骨文数字

四十 40

50 Shāng Dynasty Oracle Bone Numerals
商朝甲骨文数字

五十 50

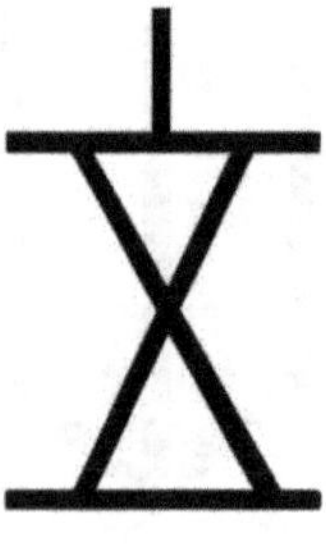

60 Shāng Dynasty Oracle Bone Numerals
商朝甲骨文数字

六十 60

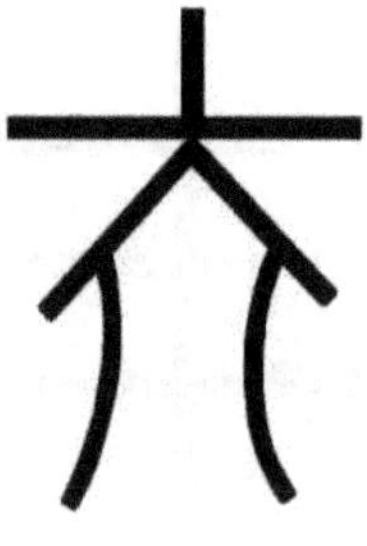

70 Shāng Dynasty Oracle Bone Numerals
商朝甲骨文数字

七十 70

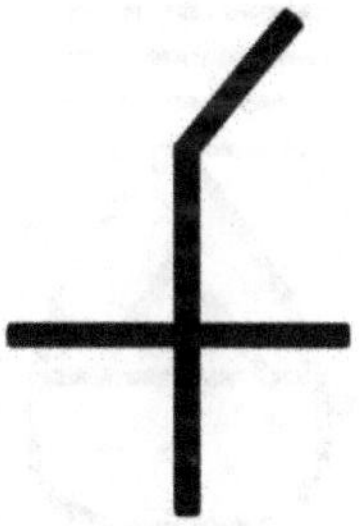

80 Shāng Dynasty Oracle Bone Numerals
商朝甲骨文数字

八十 80

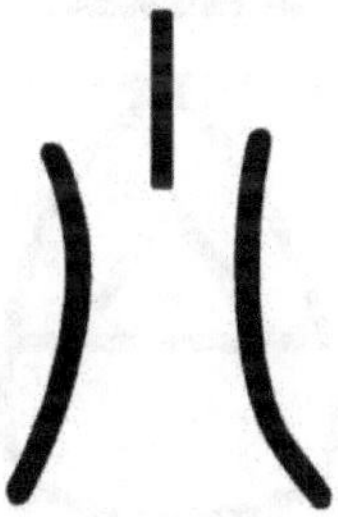

90 Shāng Dynasty Oracle Bone Numerals
商朝甲骨文数字

九十 90

100 Shāng Dynasty Oracle Bone Numerals
商朝甲骨文数字

一百 100

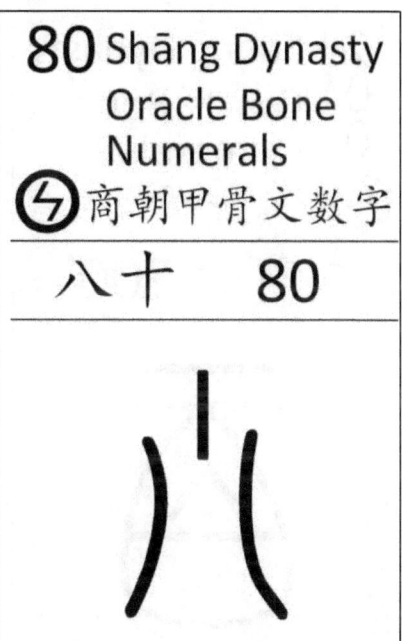

200 Shāng Dynasty
Oracle Bone
Numerals
商朝甲骨文数字

二百 200

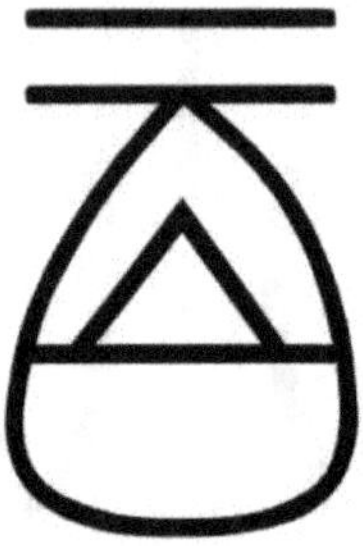

400 Shāng Dynasty
Oracle Bone
Numerals
商朝甲骨文数字

四百 400

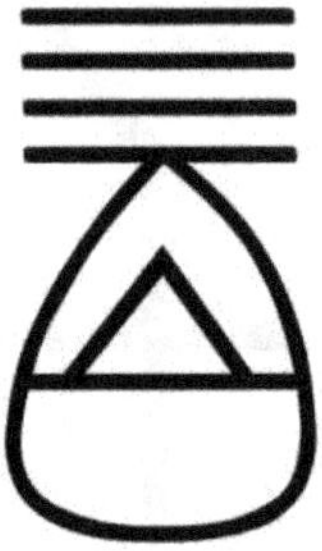

500 Shāng Dynasty
Oracle Bone
Numerals
商朝甲骨文数字

五百 500

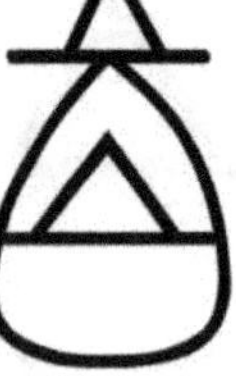

600 Shāng Dynasty
Oracle Bone
Numerals
商朝甲骨文数字

六百 600

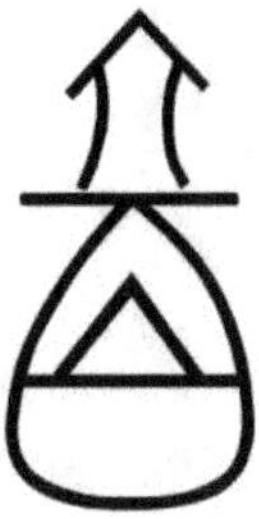

1,000

Shāng Dynasty Oracle Bone Numerals

商朝甲骨文数字

千 1,000

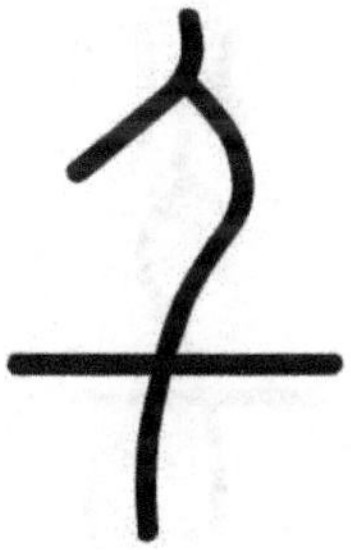

2,000

Shāng Dynasty Oracle Bone Numerals

商朝甲骨文数字

二千 2,000

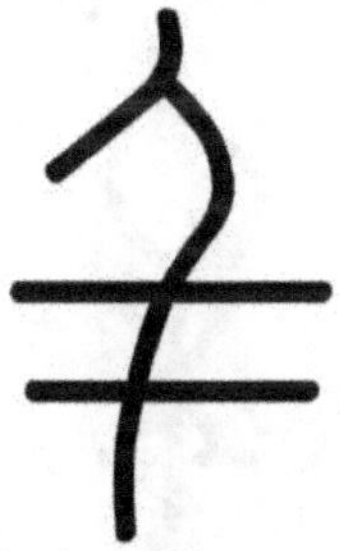

3,000

Shāng Dynasty Oracle Bone Numerals

商朝甲骨文数字

三千 3,000

4,000

Shāng Dynasty Oracle Bone Numerals

商朝甲骨文数字

四千 4,000

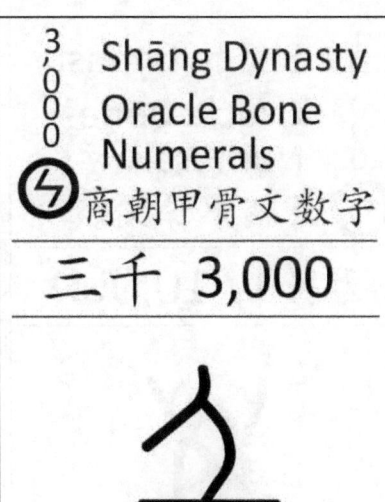

5,000
Shāng Dynasty
Oracle Bone
Numerals
商朝甲骨文数字

五千 5,000

8,000
Shāng Dynasty
Oracle Bone
Numerals
商朝甲骨文数字

八千 8,000

10,000
Shāng Dynasty
Oracle Bone
Numerals
商朝甲骨文数字

万/萬 10,000

10,000
Shāng Dynasty
Oracle Bone
Numerals
商朝甲骨文数字

万/萬 10,000

10,000 Shāng Dynasty Oracle Bone Numerals
商朝甲骨文数字

万/萬 10,000

30,000 Shāng Dynasty Oracle Bone Numerals
商朝甲骨文数字

三万/萬 30,000

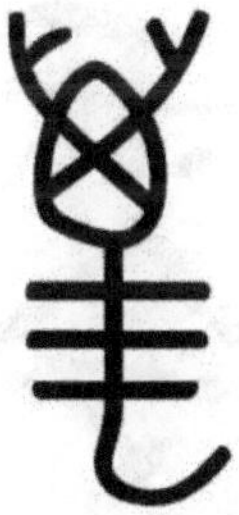

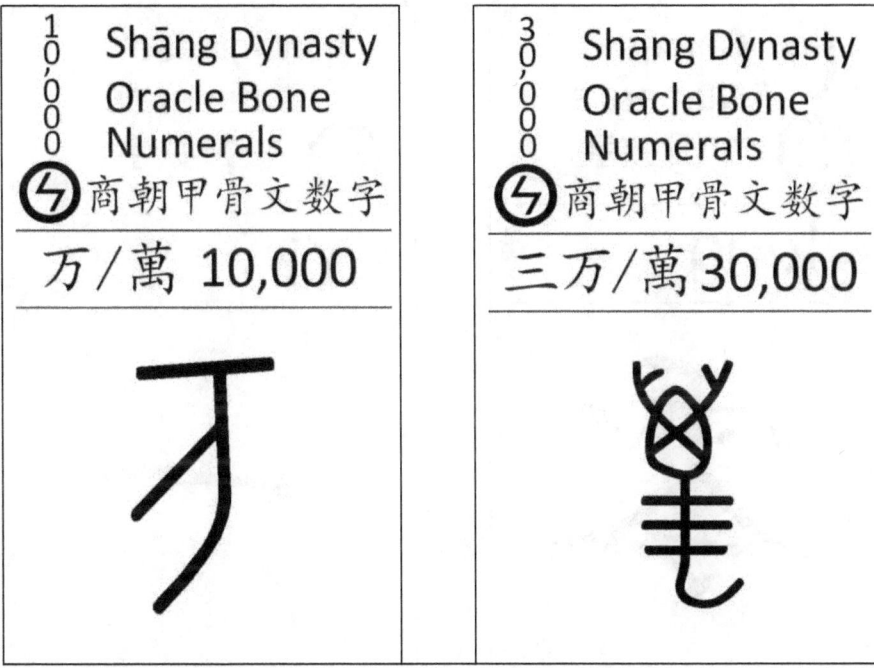

Modern Chinese Numerals 现代中国数字

Modern Chinese characters evolved slowly through the centuries and mostly have their origins in the ancient oracle bone script of the Late Shang Dynasty. A standard, regular script known today as Hàn Zì 汉子 became dominant during the Northern and Southern Dynasties period 南北朝 (420-587 AD), the Tang Dynasty calligrapher Ōuyáng Xún 欧阳询 (557-641) being its preeminent exponent. The advantages of character simplification had been discussed in China since the beginning of the 20th century and the People's Republic of China first introduced simplifications in 1956 and again in 1964. The characters for 10,000 and 100,000,000 displayed here are simplified characters. Full forms are displayed in the listing of complex numerals (大写数字 dà xiě shù zì).

0 Chinese Numerals
© 中国数字

Líng

零

1 Chinese Numerals
© 中国数字

Yī

一

2 Chinese Numerals
© 中国数字

Èr

二

3 Chinese Numerals
© 中国数字

Sān

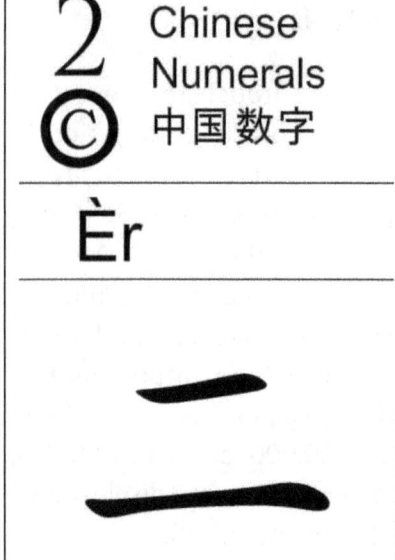

三

4 Chinese Numerals
© 中国数字

Sì

四

5 Chinese Numerals
© 中国数字

Wǔ

五

6 Chinese Numerals
© 中国数字

Liù

六

7 Chinese Numerals
© 中国数字

Qī

七

8 ©

Chinese Numerals 中国数字

Bā

八

9 ©

Chinese Numerals 中国数字

Jiŭ

九

10 ©

Chinese Numerals 中国数字

Shí

十

11 ©

Chinese Numerals 中国数字

Shí Yī

十一

12 © Chinese Numerals 中国数字	13 © Chinese Numerals 中国数字
Shí Èr	Shí Sān
十二	十三

14 © Chinese Numerals 中国数字	15 © Chinese Numerals 中国数字
Shí Sì	Shí Wǔ
十四	十五

16 Chinese Numerals
© 中国数字

Shí Liù

十六

17 Chinese Numerals
© 中国数字

Shí Qī

十七

18 Chinese Numerals
© 中国数字

Shí Bā

十八

19 Chinese Numerals
© 中国数字

Shí Jiǔ

十九

20
©
Chinese
Numerals
中国数字

Èr Shí

二十

30
©
Chinese
Numerals
中国数字

Sān Shí

三十

40
©
Chinese
Numerals
中国数字

Sì Shí

四十

50
©
Chinese
Numerals
中国数字

Wǔ Shí

五十

60 ©
Chinese
Numerals
中国数字

Liù Shí

六十

70 ©
Chinese
Numerals
中国数字

Qī Shí

七十

80 ©
Chinese
Numerals
中国数字

Bā Shí

八十

90 ©
Chinese
Numerals
中国数字

Jiǔ Shí

九十

100 Chinese Numerals
© 中国数字

Yī Bái

一百

200 Chinese Numerals
© 中国数字

Èr Bái

二百

300 Chinese Numerals
© 中国数字

Sān Bái

三百

400 Chinese Numerals
© 中国数字

Sì Bái

四百

500 Chinese Numerals © 中国数字
Wǔ Bái
五百

600 Chinese Numerals © 中国数字
Liù Bái
六百

700 Chinese Numerals © 中国数字
Qī Bái
七百

800 Chinese Numerals © 中国数字
Bā Bái
八百

900 Chinese Numerals ©
中国数字

Jiǔ Bái

九百

1,000 Chinese Numerals ©
中国数字

Yī Qiān

一千

2,000 Chinese Numerals ©
中国数字

Èr Qiān

二千

3,000 Chinese Numerals ©
中国数字

Sān Qiān

三千

4,000 Chinese Numerals ©
中国数字

Sì Qiān

四千

5,000 Chinese Numerals ©
中国数字

Wǔ Qiān

五千

6,000 Chinese Numerals ©
中国数字

Liù Qiān

六千

7,000 Chinese Numerals ©
中国数字

Qī Qiān

七千

8,000 Chinese Numerals
© 中国数字

Bā Qiān

八千

9,000 Chinese Numerals
© 中国数字

Jiǔ Qiān

九千

10,000 Chinese Numerals
© 中国数字

Yī Wàn

一万

100,000 Chinese Numerals
© 中国数字

Shí Wàn

十万

1,000,000 Chinese Numerals
©
中国数字

Yī Bái Wàn

一百万

10,000,000 Chinese Numerals
©
中国数字

Yī Qiān Wàn

一千万

100,000,000 Chinese Numerals
©
中国数字

Yī Yì

一亿

Chinese Complex Numerals 中国大写数字

Chinese complex numerals 大写数字 (dà xiě shù zì) are also known as banker's numerals or banker's anti-fraud numerals. The numerals for everyday use are too simple and can easily be changed by dishonest people intent on fraud. For example one 一 could be changed to ten 十 and two 二 could be changed to three 三 by simply adding one stroke while additional strokes could change three 三 into five 五. Complex numerals and multi-stroke un-simplified characters are far more difficult to change and their use is equivalent to the English practice of spelling out the numbers on cheques and in financial transactions.

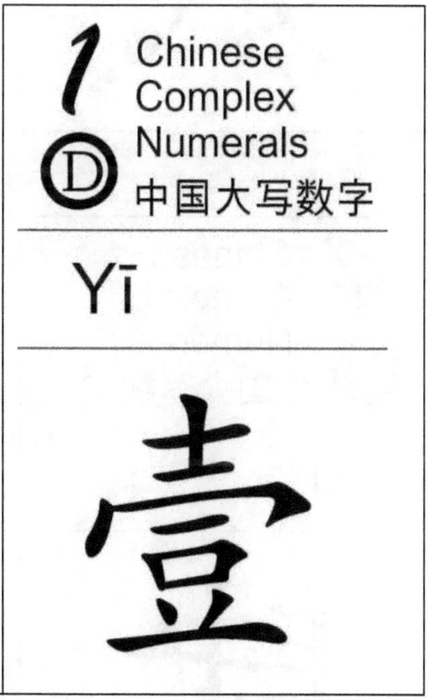

2 Chinese Complex Numerals

Ⓓ 中国大写数字

Èr

贰

3 Chinese Complex Numerals

Ⓓ 中国大写数字

Sān

叁

4 Chinese Complex Numerals

Ⓓ 中国大写数字

Sì

肆

5 Chinese Complex Numerals

Ⓓ 中国大写数字

Wǔ

伍

6 Chinese Complex Numerals
Ⓓ 中国大写数字

Liù

陆

7 Chinese Complex Numerals
Ⓓ 中国大写数字

Qī

柒

8 Chinese Complex Numerals
Ⓓ 中国大写数字

Bā

捌

9 Chinese Complex Numerals
Ⓓ 中国大写数字

Jiǔ

玖

10 Chinese Complex Numerals
D 中国大写数字

Shí

拾

11 Chinese Complex Numerals
D 中国大写数字

Shí Yī

拾壹

12 Chinese Complex Numerals
D 中国大写数字

Shí Èr

拾贰

13 Chinese Complex Numerals
D 中国大写数字

Shí Sān

拾叁

14 Chinese Complex Numerals

(D) 中国大写数字

Shí Sì

拾肆

15 Chinese Complex Numerals

(D) 中国大写数字

Shí Wǔ

拾伍

16 Chinese Complex Numerals

(D) 中国大写数字

Shí Liù

拾陆

17 Chinese Complex Numerals

(D) 中国大写数字

Shí Qī

拾柒

18 Chinese Complex Numerals
Ⓓ 中国大写数字

Shí Bā

拾捌

19 Chinese Complex Numerals
Ⓓ 中国大写数字

Shí Jiǔ

拾玖

20 Chinese Complex Numerals
Ⓓ 中国大写数字

Èr Shí

贰拾

30 Chinese Complex Numerals
Ⓓ 中国大写数字

Sān Shí

叁拾

40 Chinese Complex Numerals
Ⓓ 中国大写数字

Sì Shí

肆拾

50 Chinese Complex Numerals
Ⓓ 中国大写数字

Wǔ Shí

伍拾

60 Chinese Complex Numerals
Ⓓ 中国大写数字

Liù Shí

陆拾

70 Chinese Complex Numerals
Ⓓ 中国大写数字

Qī Shí

柒拾

80 Chinese Complex Numerals Ⓓ 中国大写数字	90 Chinese Complex Numerals Ⓓ 中国大写数字
Bā Shí	Jiǔ Shí
捌拾	玖拾

100 Chinese Complex Numerals Ⓓ 中国大写数字	200 Chinese Complex Numerals Ⓓ 中国大写数字
Yī Bái	Èr Bái
壹佰	贰佰

300 Chinese Complex Numerals
Ⓓ 中国大写数字

Sān Bái

叁佰

400 Chinese Complex Numerals
Ⓓ 中国大写数字

Sì Bái

肆佰

500 Chinese Complex Numerals
Ⓓ 中国大写数字

Wǔ Bái

伍佰

600 Chinese Complex Numerals
Ⓓ 中国大写数字

Liù Bái

陆佰

700 Chinese Complex Numerals

中国大写数字

Qī Bái

柒佰

800 Chinese Complex Numerals

中国大写数字

Bā Bái

捌佰

900 Chinese Complex Numerals

中国大写数字

Jiǔ Bái

玖佰

1,000 Chinese Complex Numerals

中国大写数字

Yī Qiān

壹仟

2,000 Chinese Complex Numerals

中国大写数字

Èr Qiān

贰仟

3,000 Chinese Complex Numerals

中国大写数字

Sān Qiān

叁仟

4,000 Chinese Complex Numerals

中国大写数字

Sì Qiān

肆仟

5,000 Chinese Complex Numerals

中国大写数字

Wǔ Qiān

伍仟

6,000 Chinese Complex Numerals

Ⓓ 中国大写数字

Liù Qiān

陆仟

7,000 Chinese Complex Numerals

Ⓓ 中国大写数字

Qī Qiān

柒仟

8,000 Chinese Complex Numerals

Ⓓ 中国大写数字

Bā Qiān

捌仟

9,000 Chinese Complex Numerals

Ⓓ 中国大写数字

Jiǔ Qiān

玖仟

10,000 Chinese Complex Numerals

Ⓓ 中国大写数字

Yī Wàn

壹萬

100,000 Chinese Complex Numerals

Ⓓ 中国大写数字

Shí Wàn

拾萬

1,000,000 Chinese Complex Numerals

Ⓓ 中国大写数字

Yī Bái Wàn

壹佰萬

10,000,000 Chinese Complex Numerals

Ⓓ 中国大写数字

Yī Qiān Wàn

壹仟萬

100,000,000 Chinese Complex Numerals

Ⓓ 中国大写数字

Yī Yì

壹億

Chinese Rod Numerals 筹算计数

In China the use of counting rods placed in the rows and columns of a counting board began in the 4th Century BC. The column farthest to the right was for units, the next column to the left for tens, followed by a column for the hundreds, a column for the thousands and so on.

Units, Hundreds, Ten Thousands, Millions, Hundred Millions etc.

𝍠	𝍡	𝍢	𝍣	𝍤	𝍥	𝍦	𝍧	𝍨
1	2	3	4	5	6	7	8	9
100	200	300	400	500	600	700	800	900
10,000	20,000	30,000	40,000	50,000	60,000	70,000	80,000	90,000

Tens, Thousands, Hundred Thousands, Ten Millions etc.

10	20	30	40	50	60	70	80	90
1,000	2,000	3,000	4,000	5,000	6,000	7,000	8,000	9,000
100,000	200,000	300,000	400,000	500,000	600,000	700,000	800,000	900,000

Small rods made of wood, bamboo or ivory were placed in the columns. The units 1-5 were represented by vertically placed rods. Six was a horizontal rod placed above a vertical rod and seven to nine were represented by additional vertical rods placed below a horizontal rod. The next column on the left was for ten which was a single horizontal rod and twenty was two horizontal rods. Additional horizontal rods were added for numbers from 30 to 50. Sixty was a vertical rod above one horizontal rod and for 70, 80 and 90 additional horizontal rods were placed below a single vertical rod, the latter progressively shortened so that all of the printed digits were the same height. One hundred, in the third column from the right, was a single vertical rod, 200 being two vertical rods etc. Then 1,000 in the 4th column from the right was a single horizontal rod and so on alternating between vertical and horizontal. The printed and written symbols of this positional numbering system were called rod numerals. These are three examples of how they are written.

162,149 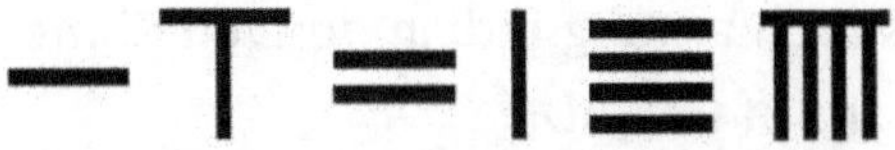

530,588 

8,738, 907,460

By about 50 AD some Chinese mathematicians began to use negative numbers which were represented by black rods on a counting board while positive numbers were indicated by red rods. At that time all counting rods had round cylindrical shapes but by the time of the Suí Dynasty 隋朝 (581-618) counting rods for positive numbers were triangular while negative rods were rectangular. Zero was at first a blank space but the Mathematical Treatise in Nine Sections (数书九章 Shùshū Jiǔzhāng) of 1247 by Qín Jiǔsháo 秦九韶 uses a circle for zero. This may have evolved from the concept of Wújí 无极 (無極) in Daoism which is represented by a circle. Wújí is the limitless cosmic first principle which predates the existential and material world and therefore represents nothingness. It is followed by Tàijí 太极 the great primal beginning which is portrayed as a circle divided into dark and light sections known as Yīn 阴 and Yáng 阳. Tàijí is thus the Yīn-Yáng principle of bipolarity from which existence evolved.

Wújí 无极  Tàijí 太极 

However, some historians think that the Indian monk Gautama Siddha took the circle as a sign for zero to China in 718. On the other hand, zero was at first represented by a dot in India and it is clearly shown as such among Indian derived Khmer numerals in the Sambor inscriptions of 683 AD.

The number 605 in the Sambor inscription 

A circle was among the Chinese characters promoted by Empress Wǔ Zétiān 武则天 in 690 AD. Then a stone inscription dated 876 at the Chaturbhuja Temple at Gwalior, India also shows a circle as a symbol for zero.

The use of negative numbers in China by 50 AD, the circular symbol in Daoism and the promotion of the circle by Wŭ Zétiān 武则天 in 690, all suggest that the circle as a symbol of zero may have evolved in China. There may have been a Chinese circular zero long before Qín Jiŭsháo used it in 1247 although evidence that would prove this has yet to be found. Based on Rod Calculus described in *The Mathematical Classic of Master Sun* 孙子算经 *Sūnzĭ Suànjīng* (3rd to 5th Century AD), the Singaporean mathematics historian Lam Lay Yong (蓝丽蓉 Lán Lìróng); claims that the Hindu-Arabic numeral system had its origin in China.

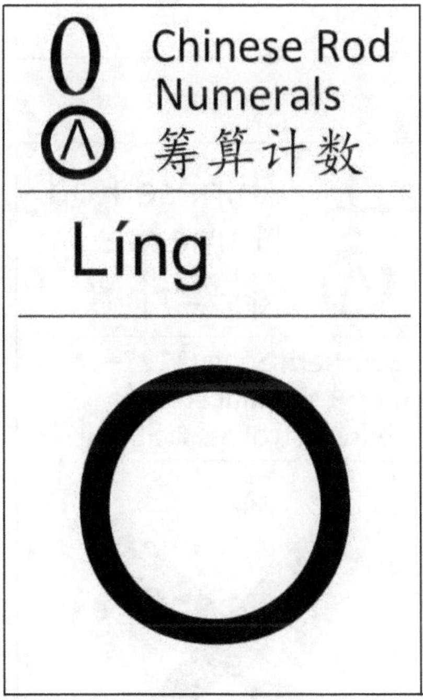

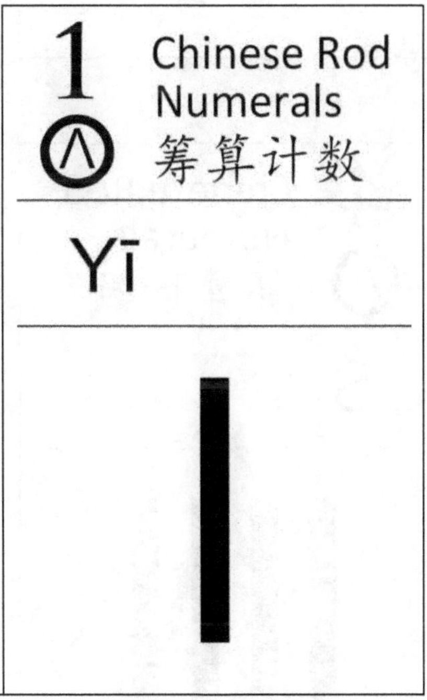

2 Chinese Rod Numerals 筹算计数

Èr

3 Chinese Rod Numerals 筹算计数

Sān

4 Chinese Rod Numerals 筹算计数

Sì

4 Chinese Rod Numerals 筹算计数

Southern Sòng (1127–1279) simplification to reduce strokes 南宋简化版

5 Chinese Rod Numerals 筹算计数

Wǔ

5 Chinese Rod Numerals 筹算计数

Southern Sòng (1127–1279) simplification to reduce strokes 南宋简化版

6 Chinese Rod Numerals 筹算计数

Liù

7 Chinese Rod Numerals 筹算计数

Qī

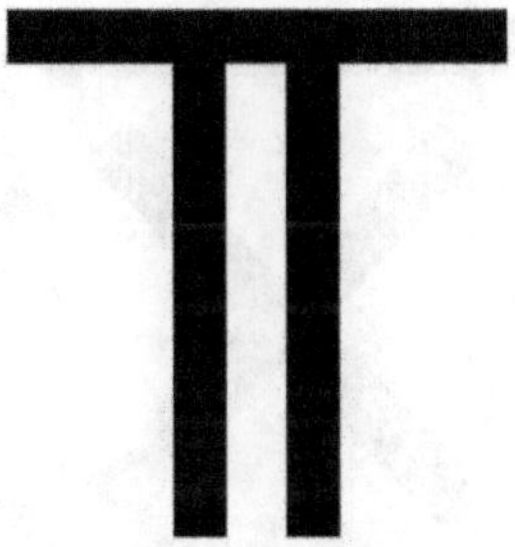

8 Chinese Rod Numerals 筹算计数

Bā

9 Chinese Rod Numerals 筹算计数

Jiŭ

9 Chinese Rod Numerals 筹算计数

Southern Sòng (1127–1279) simplification to reduce strokes 南宋简化版

10 Chinese Rod Numerals 筹算计数

Shí

11 Chinese Rod Numerals
筹算计数

Shí Yī

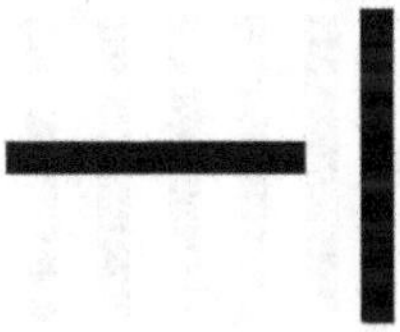

12 Chinese Rod Numerals
筹算计数

Shí Èr

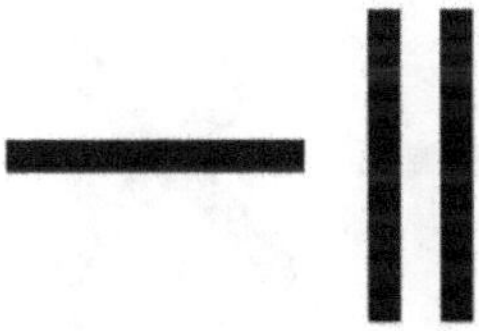

13 Chinese Rod Numerals
筹算计数

Shí Sān

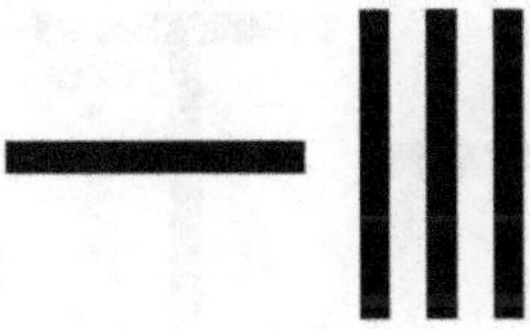

14 Chinese Rod Numerals
筹算计数

Shí Sì

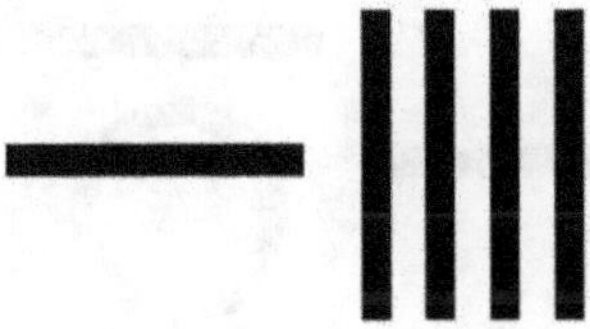

14 Chinese Rod Numerals
筹算计数

Southern Sòng (1127–1279) simplification to reduce strokes 南宋简化版

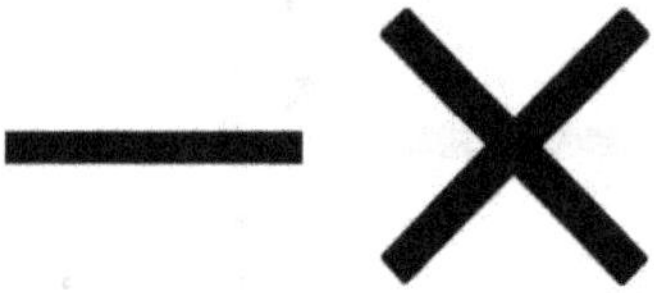

15 Chinese Rod Numerals
筹算计数

Shí Wǔ

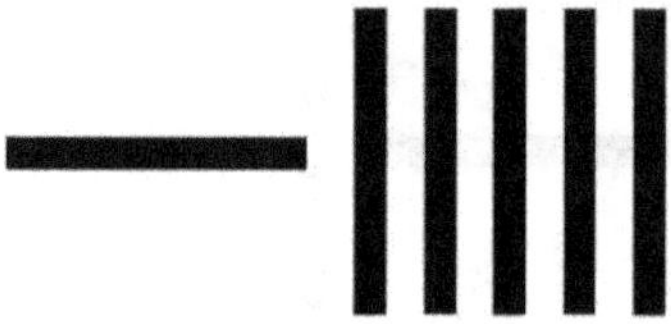

15 Chinese Rod Numerals
筹算计数

Southern Sòng (1127–1279) simplification to reduce strokes 南宋简化版

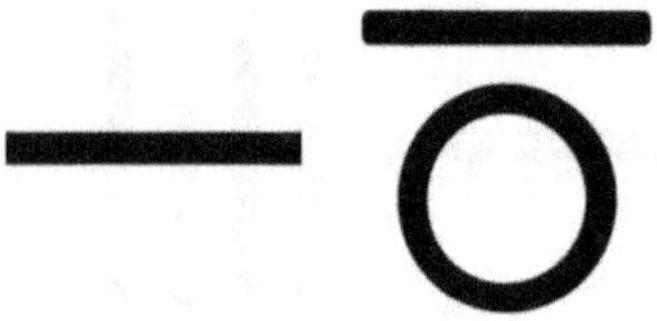

16 Chinese Rod Numerals
筹算计数

Shí Liù

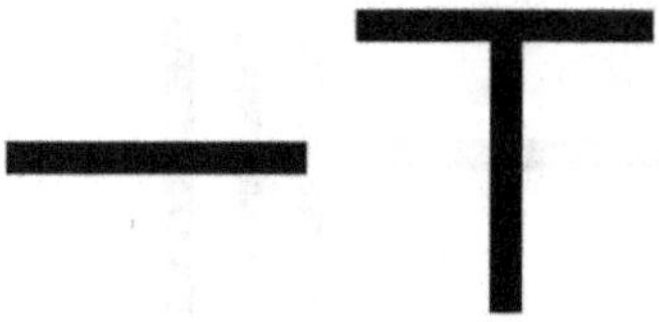

17 Chinese Rod Numerals 筹算计数

Shí Qī

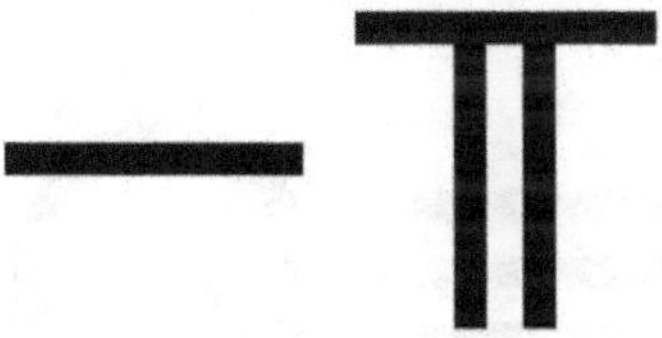

18 Chinese Rod Numerals 筹算计数

Shí Bā

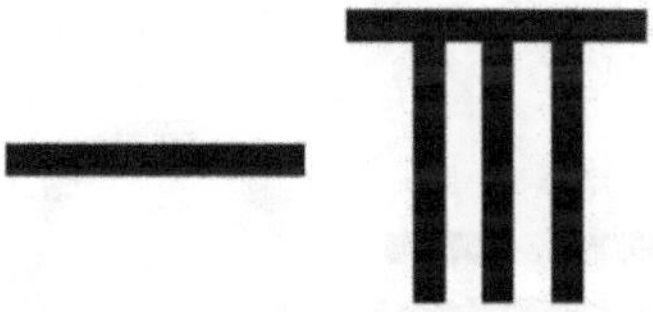

19 Chinese Rod Numerals 筹算计数

Shí Jiǔ

19 Chinese Rod Numerals 筹算计数

Southern Sòng (1127–1279) simplification to reduce strokes 南宋简化版

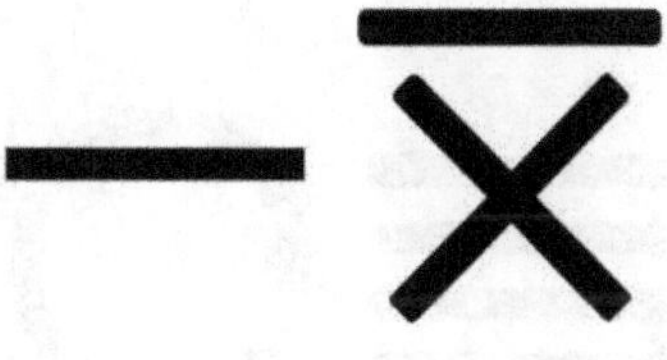

20 Chinese Rod Numerals 筹算计数

Èr Shí

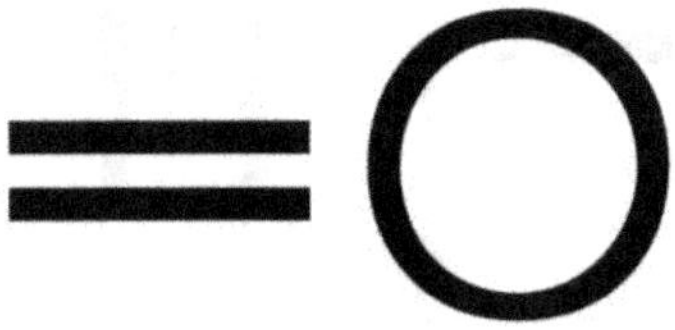

30 Chinese Rod Numerals 筹算计数

Sān Shí

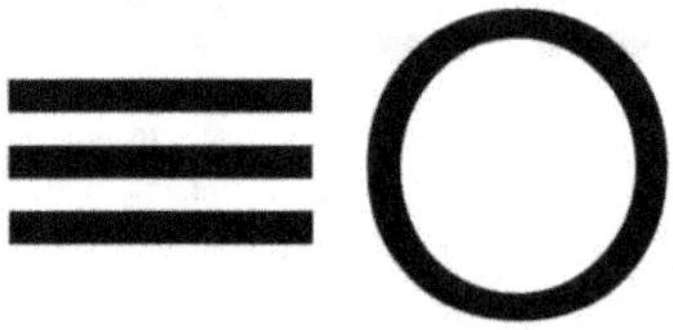

40 Chinese Rod Numerals 筹算计数

Sì Shí

50 Chinese Rod Numerals 筹算计数

Wǔ Shí

50 Chinese Rod Numerals 筹算计数

Southern Sòng (1127–1279) simplification to reduce strokes 南宋简化版

60 Chinese Rod Numerals 筹算计数

Liù Shí

70 Chinese Rod Numerals 筹算计数

Qī Shí

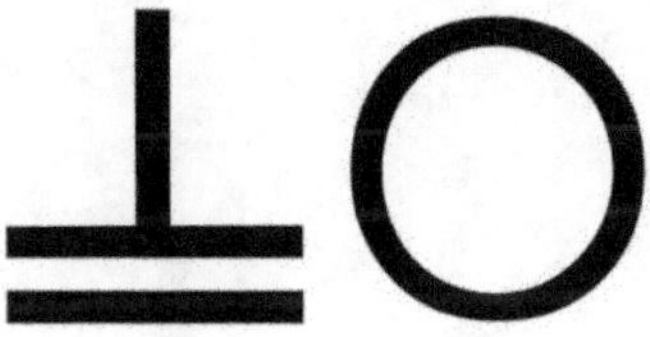

80 Chinese Rod Numerals 筹算计数

Bā Shí

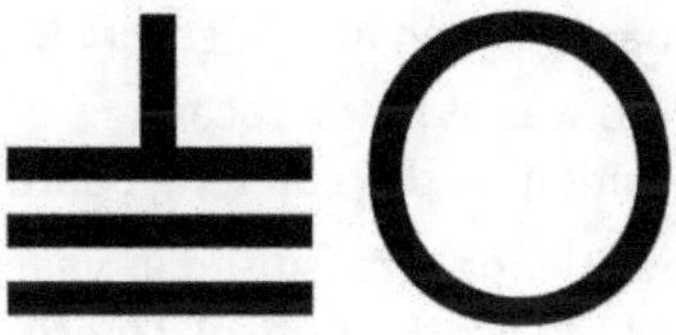

90 Chinese Rod Numerals

筹算计数

Jiǔ Shí

90 Chinese Rod Numerals

筹算计数

Southern Sòng (1127–1279) simplification to reduce strokes 南宋简化版

Chinese Rod Numerals 382,670,495

筹算计数

382,670,495

Rod numerals used on counting boards began in 4th Century BC.

Zero was at first a vacant position on the counting board.

A Circle for zero is used in year 1247 in ***Mathematical Treatise in Nine Sections***: 数书九章 by Qín Jiǔsháo 秦九韶.

Negative Chinese Rod Numerals 筹算计数一负数

By about 50 AD some Chinese mathematicians began to use negative numbers which were represented by black rods on a counting board while positive numbers were indicated by red rods. At that time all counting rods had round cylindrical shapes but by the time of the Suí Dynasty 隋朝 (581-618) counting rods for positive numbers were triangular while negative rods were rectangular. Zero was at first a blank space but the Mathematical Treatise in Nine Sections (数书九章 Shùshū Jiǔzhāng) of 1247 by Qín Jiǔsháo 秦九韶 uses a circle for zero. Negative rod numerals are written with a slant bar across the last digit.

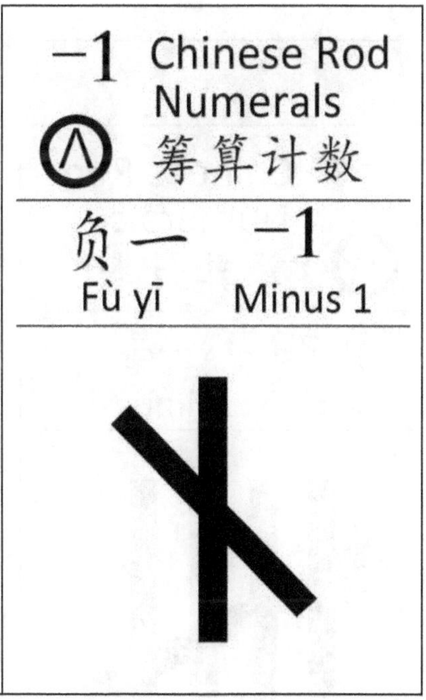

−2 Chinese Rod Numerals
筹算计数

负二 −2
Fù èr Minus 2

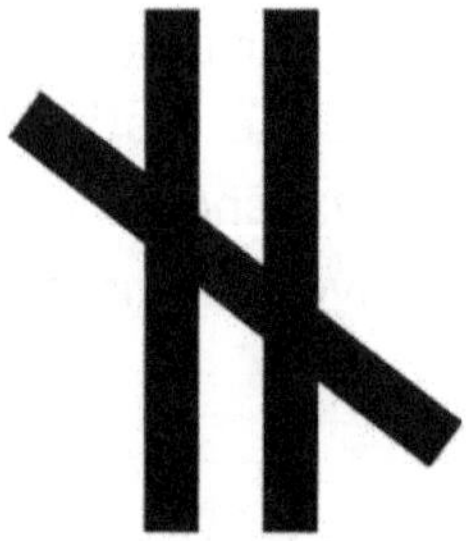

−3 Chinese Rod Numerals
筹算计数

负三 −3
Fù sān Minus 3

−4 Chinese Rod Numerals
筹算计数

负四 −4
Fù sì Minus 4

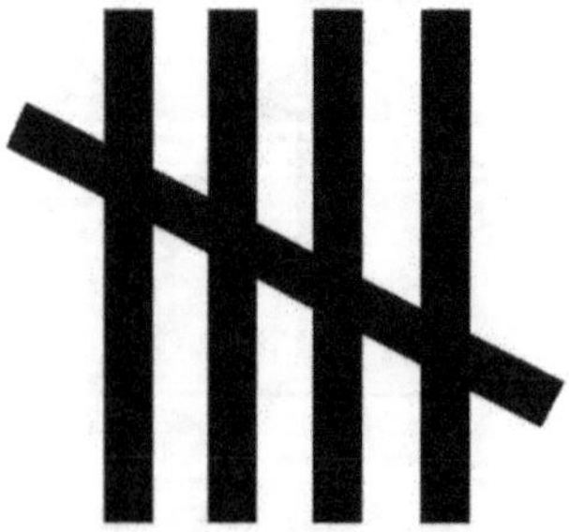

−5 Chinese Rod Numerals
筹算计数

负五 −5
Fù wǔ Minus 5

−6 Chinese Rod Numerals
筹算计数

负六 −6
Fù liù Minus 6

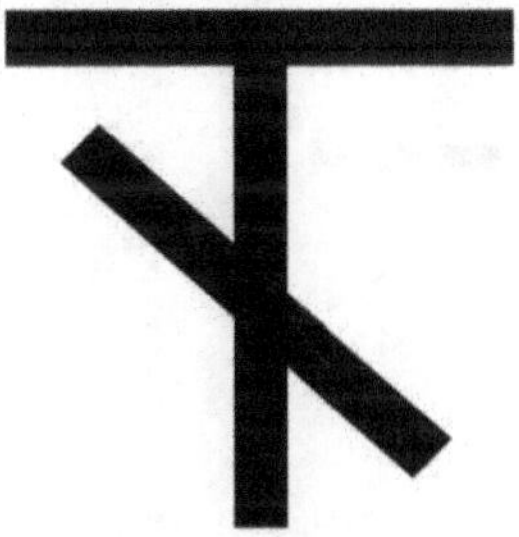

−7 Chinese Rod Numerals
筹算计数

负七 −7
Fù qī Minus 7

−8 Chinese Rod Numerals
筹算计数

负八 −8
Fù bā Minus 8

−9 Chinese Rod Numerals
筹算计数

负九 −9
Fù jiǔ Minus 9

−10 Chinese Rod Numerals

筹算计数

负十 −10

Fù shí Minus 10

−11 Chinese Rod Numerals

筹算计数

负十一 −11

Fù shí yī Minus 11

−12 Chinese Rod Numerals

筹算计数

负十二 −12

Fù shí èr Minus 12

−13 Chinese Rod Numerals

筹算计数

负十三 −13

Fù shí sān Minus 13

−14 Chinese Rod Numerals
筹算计数

负十四 −14
Fù shí sì Minus 14

−15 Chinese Rod Numerals
筹算计数

负十五 −15
Fù shí wǔ Minus 15

−16 Chinese Rod Numerals
筹算计数

负十六 −16
Fù shí liù Minus 16

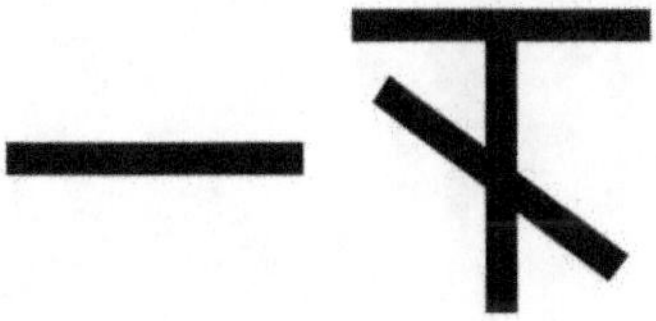

−17 Chinese Rod Numerals
筹算计数

负十七 −17
Fù shí qī Minus 17

−18 Chinese Rod Numerals

筹算计数

负十八 −18

Fù shí bā Minus 18

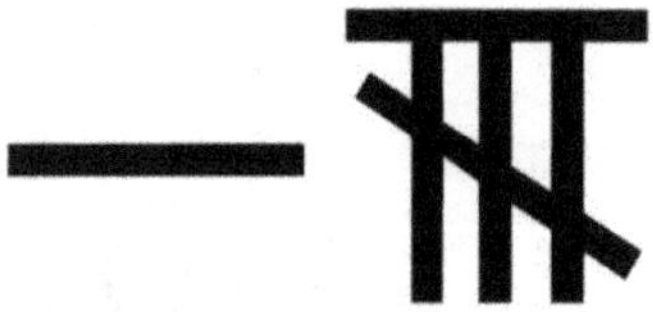

−19 Chinese Rod Numerals

筹算计数

负十九 −19

Fù shí jiǔ Minus 19

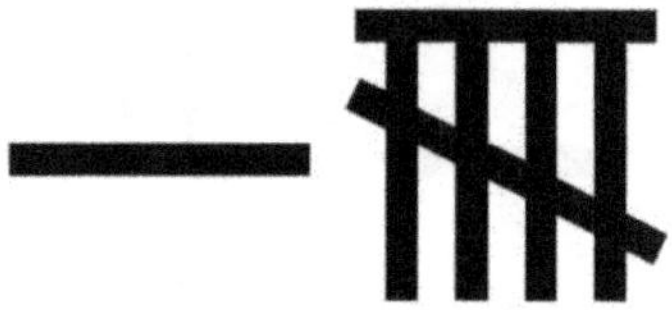

−20 Chinese Rod Numerals

筹算计数

负二十 −20

Fù èr shí Minus 20

The Suzhou Numerals 苏州花码

The Suzhou Numerals are a variation of rod numerals used during the Southern Song Dynasty (1127–1279). They may still be used in some parts of Southern China to display prices in markets and restaurants but have been largely replaced by the use of the Indo-Arabic numerals derived from Western countries. The Suzhou Numerals 5 and 9 evolved from Southern Sòng 南宋 simplifications of the corresponding horizontal rod numerals. Most of the Suzhou numerals have their origins in Chinese rod numerals or their Southern Sòng simplifications, including 20, 30 and 40 which have vertical bars placed across the horizontal rod numeral 10.

Suzhou numeral 5  evolved from

Southern Sòng simplified horizontal rod numeral 5  .

Suzhou numeral 9  evolved from

Southern Sòng simplified horizontal rod numeral 9  .

Suzhou numeral 4  comes from Southern Sòng

simplified vertical or horizontal rod numeral 4  .

0

The Sūzhōu Numerals
苏州花码

Líng

〇

1

The Sūzhōu Numerals
苏州花码

Yī

〡

2

The Sūzhōu Numerals
苏州花码

Èr

〢

3

The Sūzhōu Numerals
苏州花码

Sān

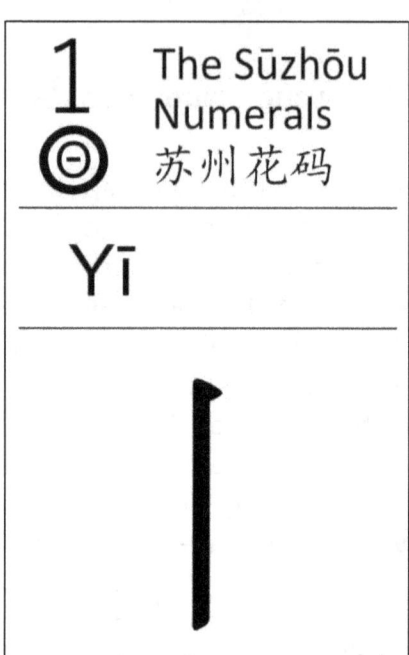

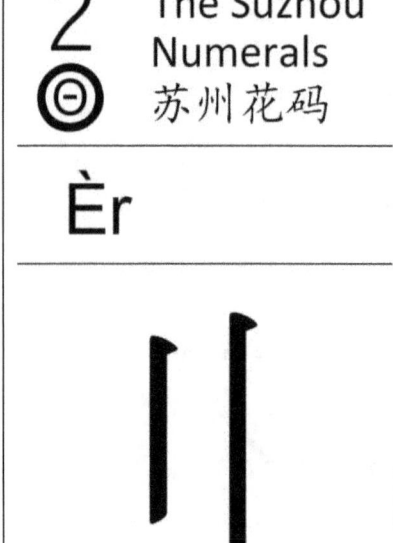

〣

4 The Sūzhōu Numerals 苏州花码

Sì

〤

5 The Sūzhōu Numerals 苏州花码

Wǔ

〥

6 The Sūzhōu Numerals 苏州花码

Liù

〦

7 The Sūzhōu Numerals 苏州花码

Qī

〧

8 The Sūzhōu Numerals 苏州花码

Bā

〨

9 The Sūzhōu Numerals 苏州花码

Jiǔ

〩

10 The Sūzhōu Numerals 苏州花码

Shí

十

11 The Sūzhōu Numerals 苏州花码

Shí Yī

十〡

12 The Sūzhōu Numerals 苏州花码

Shí Èr

〸〢

13 The Sūzhōu Numerals 苏州花码

Shí Sān

〸〣

14 The Sūzhōu Numerals 苏州花码

Shí Sì

〸〤

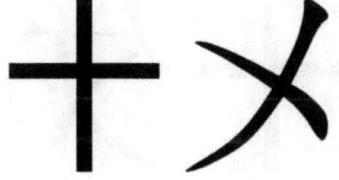

15 The Sūzhōu Numerals 苏州花码

Shí Wǔ

〸〥

16 The Sūzhōu Numerals 苏州花码

Shí Liù

十〦

17 The Sūzhōu Numerals 苏州花码

Shí Qī

十〧

18 The Sūzhōu Numerals 苏州花码

Shí Bā

十〨

19 The Sūzhōu Numerals 苏州花码

Shí Jiǔ

十〩

20 The Sūzhōu Numerals 苏州花码

Èr Shí

廿

30 The Sūzhōu Numerals 苏州花码

Sān Shí

卅

40 The Sūzhōu Numerals 苏州花码

Sì Shí

卌

50 The Sūzhōu Numerals 苏州花码

Wǔ Shí

〥十

60 The Sūzhōu Numerals 苏州花码

Liù Shí

〦十

70 The Sūzhōu Numerals 苏州花码

Qī Shí

〧十

80 The Sūzhōu Numerals 苏州花码

Bā Shí

〨十

90 The Sūzhōu Numerals 苏州花码

Jiǔ Shí

〩十

Babylonian Numerals 巴比伦楔形数字

Although writing began in Sumer by about 3,300 BC, the Babylonian wedge-shaped numeral system first appeared around 2000 BC. It was a sexagesimal (base 60) positional numbering system with a ten digit sub-base. Only two symbols were needed to count to 59, the symbol for one and the symbol for ten. The 59 digits were formed by combinations of these two symbols. A space was left for zero and at a later date a symbol for zero within a number was created but there was no symbol for a final zero. Numbers were impressed in soft clay with a wedge-shaped stylus after which the clay was left to dry and harden in the sun creating a permanent record. Elements of the Babylonian sexagesimal system still survive in our minutes of 60 seconds, our hours of 60 minutes and the equilateral triangle with angles of 60 degrees. The base-60 system and the circle of 360 degrees may derive from the Babylonian year of 360 days (6 x 60).

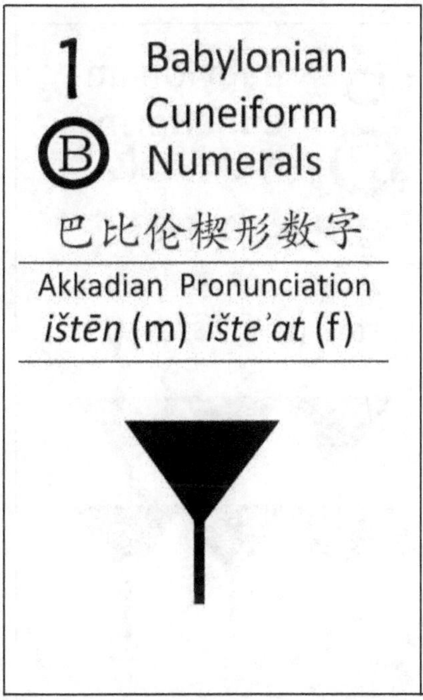

3 Babylonian Cuneiform Numerals

Ⓑ

巴比伦楔形数字

Akkadian Pronunciation

šalāš (m) *šalāšat* (f)

4 Babylonian Cuneiform Numerals

Ⓑ

巴比伦楔形数字

Akkadian Pronunciation

erbē (m) *erbēt* (f)

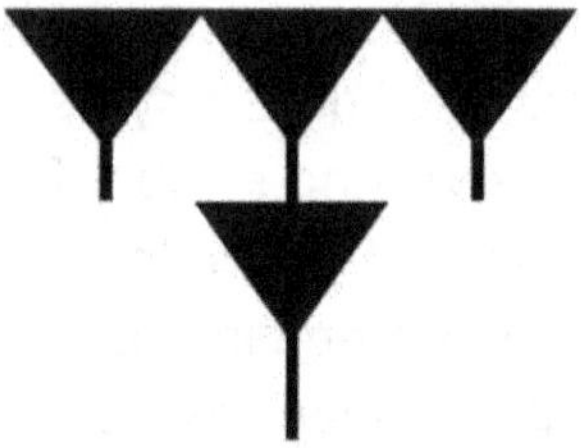

5 Babylonian Cuneiform Numerals

Ⓑ

巴比伦楔形数字

Akkadian Pronunciation

ḫamiš (m) *ḫamšat* (f)

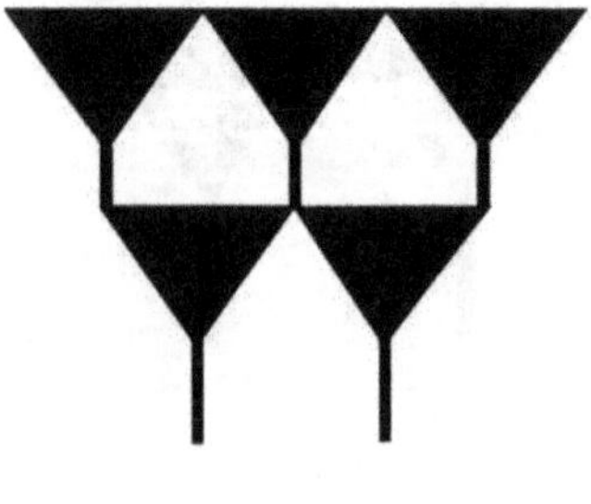

6 Babylonian Cuneiform Numerals

Ⓑ

巴比伦楔形数字

Akkadian Pronunciation

šediš (m) *šiššet* (f)

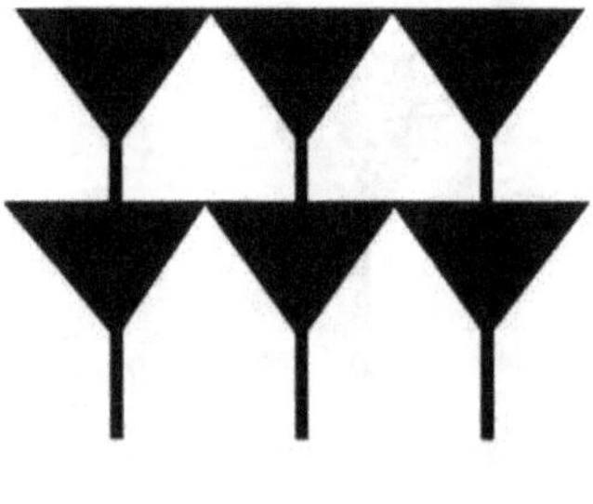

7 Ⓑ Babylonian Cuneiform Numerals

巴比伦楔形数字

Akkadian Pronunciation

sebē (m) *sebēt* (f)

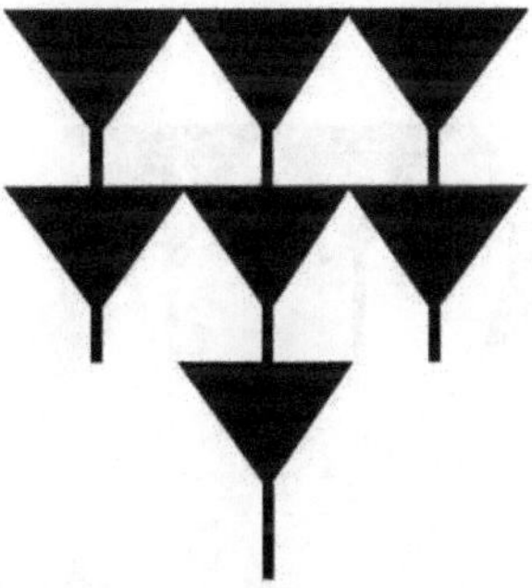

8 Ⓑ Babylonian Cuneiform Numerals

巴比伦楔形数字

Akkadian Pronunciation

samānē (m) *samānat* (f)

9 Ⓑ Babylonian Cuneiform Numerals

巴比伦楔形数字

Akkadian Pronunciation

tešē (m) *tišīt* (f)

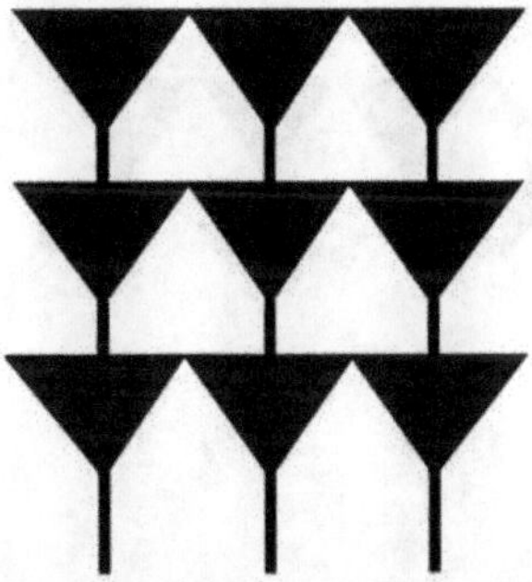

10 Ⓑ Babylonian Cuneiform Numerals

巴比伦楔形数字

Akkadian Pronunciation

ešer (m) *ešeret* (f)

11 Babylonian Cuneiform Numerals

Ⓑ

巴比伦楔形数字

𒁹 = 1   𒌋 = 10

12 Babylonian Cuneiform Numerals

Ⓑ

巴比伦楔形数字

𒁹 = 1   𒌋 = 10

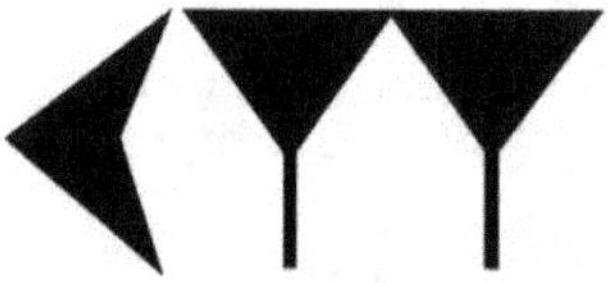

13 Babylonian Cuneiform Numerals

Ⓑ

巴比伦楔形数字

𒁹 = 1   𒌋 = 10

14 Babylonian Cuneiform Numerals

Ⓑ

巴比伦楔形数字

𒁹 = 1   𒌋 = 10

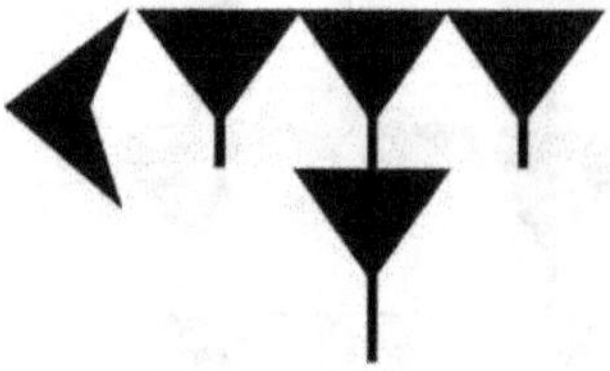

15 Babylonian Cuneiform Numerals

Ⓑ

巴比伦楔形数字

𒁹 = 1 𒌋 = 10

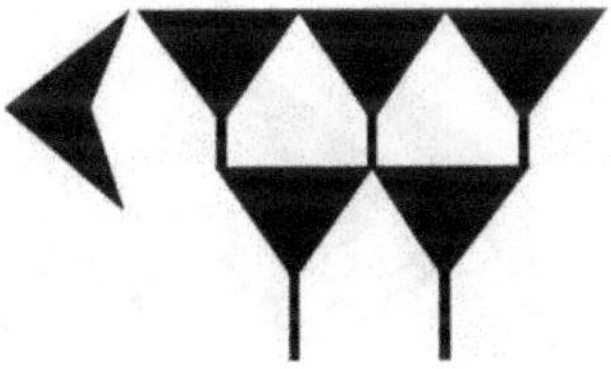

16 Babylonian Cuneiform Numerals

Ⓑ

巴比伦楔形数字

𒁹 = 1 𒌋 = 10

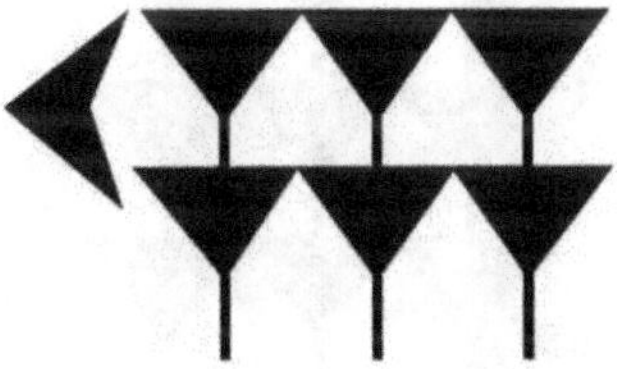

17 Babylonian Cuneiform Numerals

Ⓑ

巴比伦楔形数字

𒁹 = 1 𒌋 = 10

18 Babylonian Cuneiform Numerals

Ⓑ

巴比伦楔形数字

𒁹 = 1 𒌋 = 10

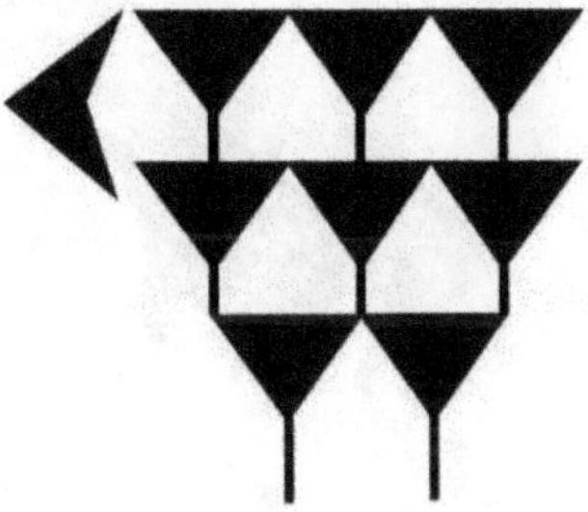

19 Babylonian Cuneiform Numerals (B)

巴比伦楔形数字

𒁹 = 1 𒌋 = 10

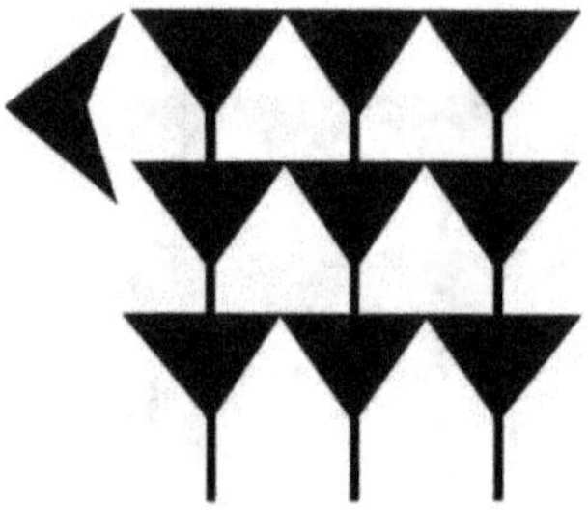

20 Babylonian Cuneiform Numerals (B)

巴比伦楔形数字

𒁹 = 1 𒌋 = 10

30 Babylonian Cuneiform Numerals (B)

巴比伦楔形数字

𒁹 = 1 𒌋 = 10

40 Babylonian Cuneiform Numerals (B)

巴比伦楔形数字

𒁹 = 1 𒌋 = 10

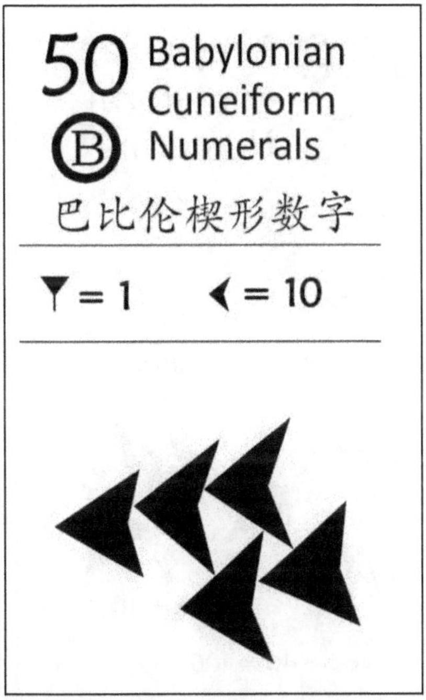
50 Babylonian Cuneiform Numerals
巴比伦楔形数字
= 1
= 10

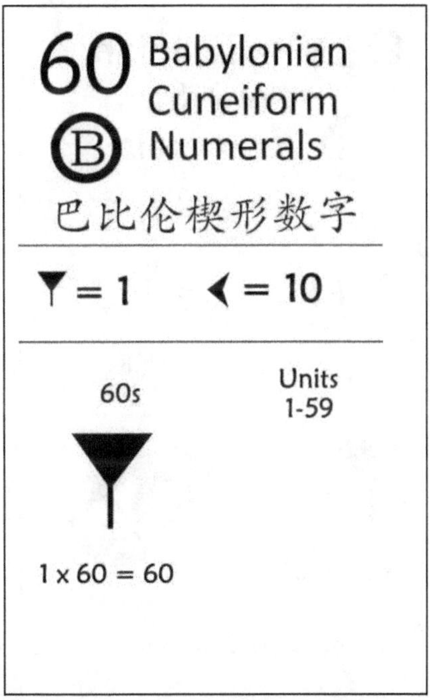
60 Babylonian Cuneiform Numerals
巴比伦楔形数字
= 1
= 10
60s
Units 1-59
1 x 60 = 60

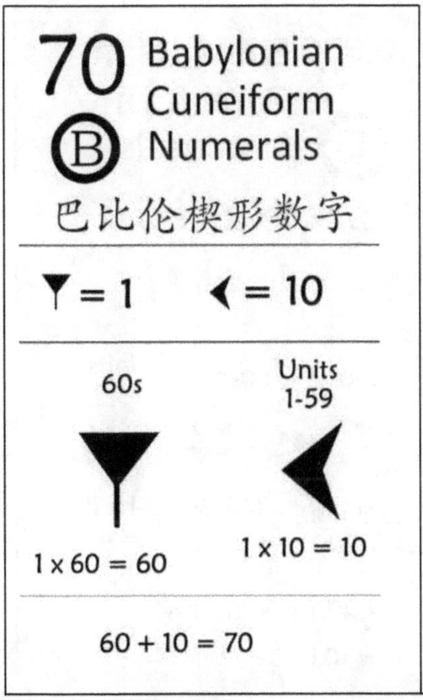
70 Babylonian Cuneiform Numerals
巴比伦楔形数字
= 1
= 10
60s
Units 1-59
1 x 60 = 60
1 x 10 = 10
60 + 10 = 70

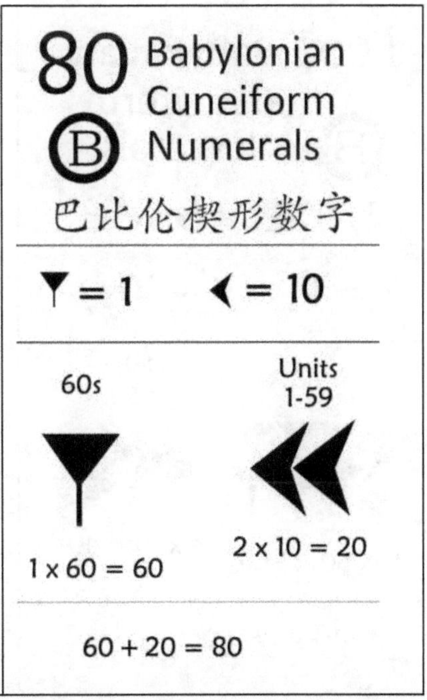
80 Babylonian Cuneiform Numerals
巴比伦楔形数字
= 1
= 10
60s
Units 1-59
1 x 60 = 60
2 x 10 = 20
60 + 20 = 80

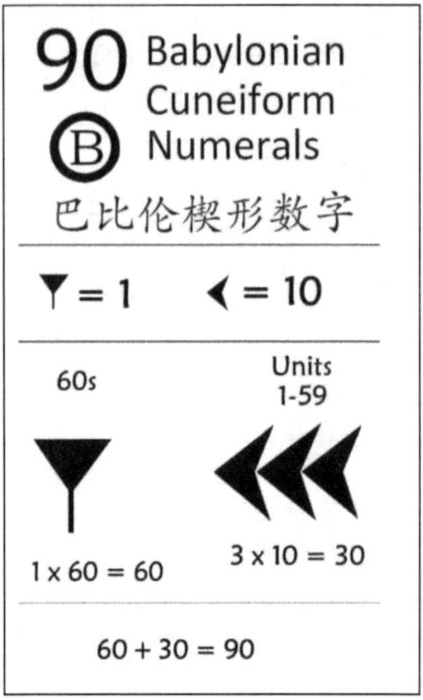
90 Babylonian Cuneiform Numerals
B
巴比伦楔形数字
= 1
= 10
60s
Units 1-59
1 x 60 = 60
3 x 10 = 30
60 + 30 = 90

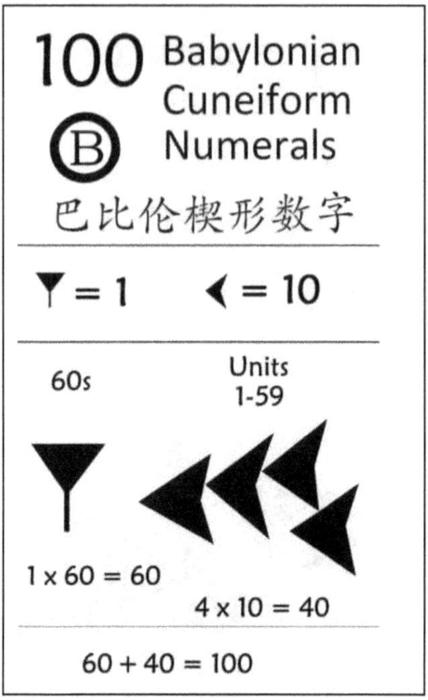
100 Babylonian Cuneiform Numerals
B
巴比伦楔形数字
= 1
= 10
60s
Units 1-59
1 x 60 = 60
4 x 10 = 40
60 + 40 = 100

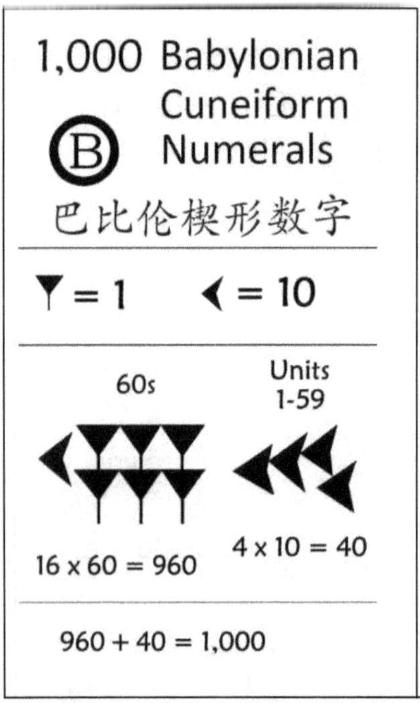
1,000 Babylonian Cuneiform Numerals
B
巴比伦楔形数字
= 1
= 10
60s
Units 1-59
16 x 60 = 960
4 x 10 = 40
960 + 40 = 1,000

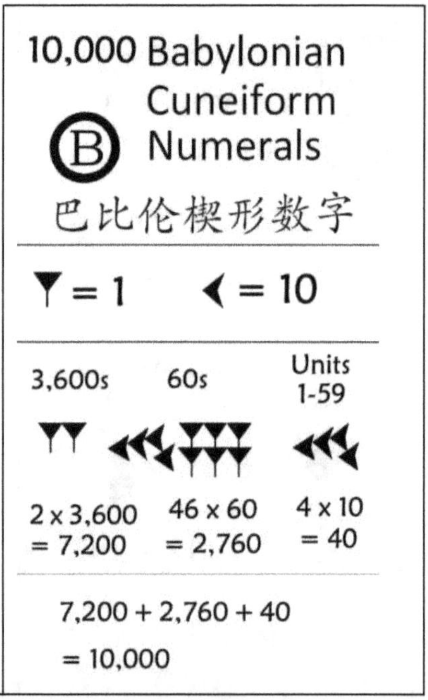
10,000 Babylonian Cuneiform Numerals
B
巴比伦楔形数字
= 1
= 10
3,600s
60s
Units 1-59
2 x 3,600 = 7,200
46 x 60 = 2,760
4 x 10 = 40
7,200 + 2,760 + 40 = 10,000

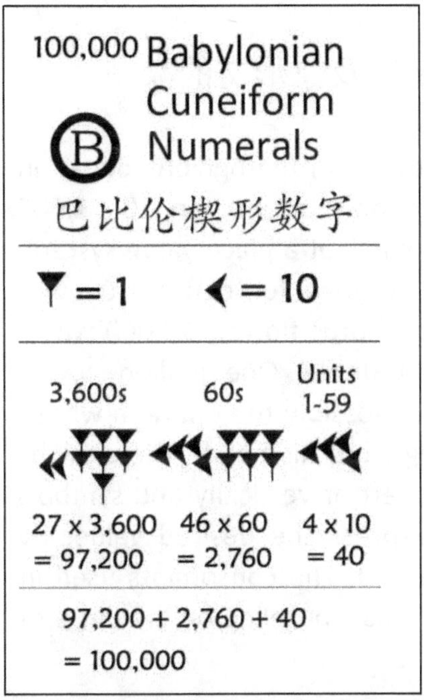
100,000 Babylonian Cuneiform Numerals
B
巴比伦楔形数字
= 1
= 10
3,600s
60s
Units 1-59
27 x 3,600 = 97,200
46 x 60 = 2,760
4 x 10 = 40
97,200 + 2,760 + 40 = 100,000

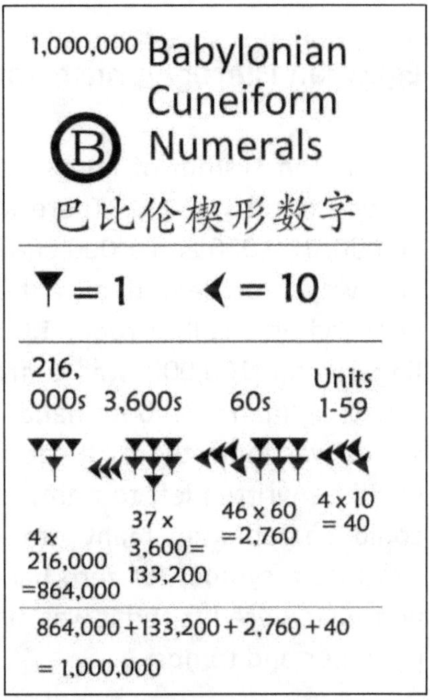
1,000,000 Babylonian Cuneiform Numerals
B
巴比伦楔形数字
= 1
= 10
216, 000s
3,600s
60s
Units 1-59
4 x 216,000 =864,000
37 x 3,600= 133,200
46 x 60 =2,760
4 x 10 = 40
864,000+133,200+2,760+40 = 1,000,000

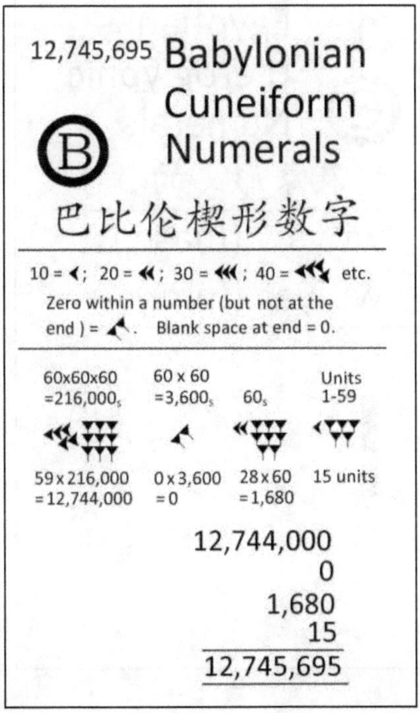
12,745,695 Babylonian Cuneiform Numerals
B
巴比伦楔形数字
10 = ; 20 = ; 30 = ; 40 = etc.
Zero within a number (but not at the end) = . Blank space at end = 0.
60x60x60 =216,000s
60 x 60 =3,600s
60s
Units 1-59
59x216,000 =12,744,000
0x3,600 =0
28x60 =1,680
15 units
12,744,000
0
1,680
15
12,745,695

Egyptian Hieroglyphic Numerals 埃及象形文的数字

A base-ten system of numerals written with hieroglyphs began in Egypt around 3,000 BC. There were symbols for one, ten, 100, 1,000, 10,000, 100,000 and 1,000,000 but it was not a place value system. Ten was represented by a hobble or yoke for cattle; 100 was depicted as a coil of rope; 1,000 was a lotus flower; 10,000 was a finger; and 100,000 was a frog or tadpole. One million was a squatting man with both hands raised, possibly the pharaoh whose status was indicated by a cobra rising from his head. Hieroglyphs could be written left to right, right to left or vertically and symbols could be repeated many times to express the desired value. By 1740 BC a symbol for zero had appeared. The consonants used in writing it were *nfr* and it was also used as a baseline on drawings of pyramids and tombs.

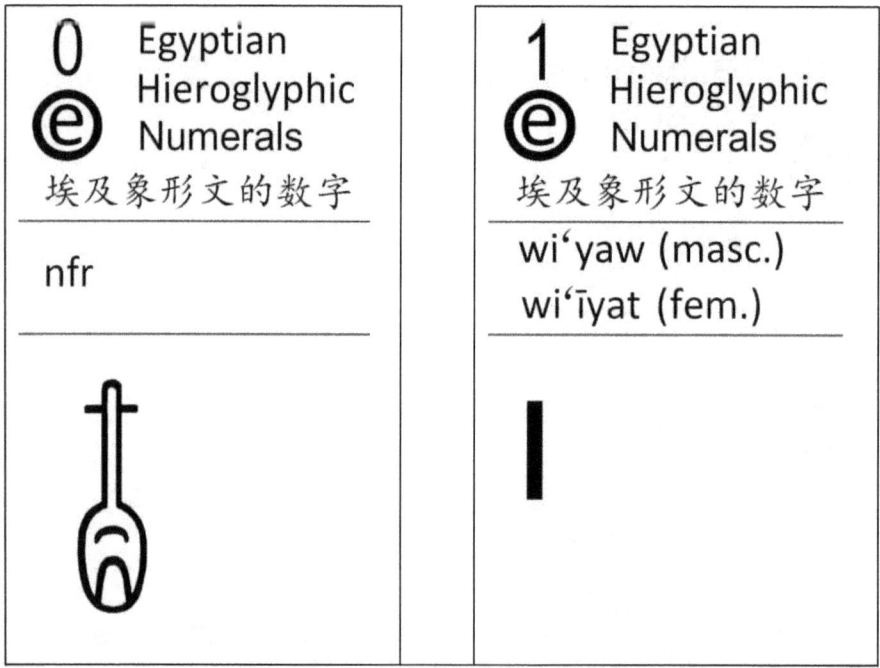

0 ⓔ Egyptian Hieroglyphic Numerals 埃及象形文的数字	1 ⓔ Egyptian Hieroglyphic Numerals 埃及象形文的数字
nfr	wiʿyaw (masc.) wiʿīyat (fem.)

2 Egyptian Hieroglyphic Numerals

埃及象形文的数字

sínway (masc.)
síntay (fem.)

3 Egyptian Hieroglyphic Numerals

埃及象形文的数字

ḫámtaw (masc.)
ḫámtat (fem.)

4 Egyptian Hieroglyphic Numerals

埃及象形文的数字

yAfdáw (masc.)
yAfdát (fem.)

5 Egyptian Hieroglyphic Numerals

埃及象形文的数字

dīyaw (masc.)
dīyat (fem.)

6 ⓔ Egyptian Hieroglyphic Numerals

埃及象形文的数字

yAssáw (masc.)
yAssát (fem.)

7 ⓔ Egyptian Hieroglyphic Numerals

埃及象形文的数字

sáfḫaw (masc.)
sáfḫat (fem.)

8 ⓔ Egyptian Hieroglyphic Numerals

埃及象形文的数字

ḫAmānaw (masc.)
ḫAmānat (fem.)

9 ⓔ Egyptian Hieroglyphic Numerals

埃及象形文的数字

pAsīḏaw (masc.)
pAsīḏat (fem.)

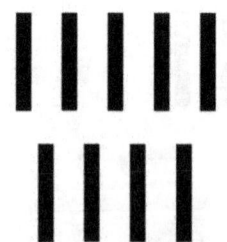

10 Egyptian Hieroglyphic Numerals
埃及象形文的数字

mūḏaw (masc.)
mūḏat (fem.)

11 Egyptian Hieroglyphic Numerals
埃及象形文的数字

11

12 Egyptian Hieroglyphic Numerals
埃及象形文的数字

12

13 Egyptian Hieroglyphic Numerals
埃及象形文的数字

13

14 Egyptian Hieroglyphic Numerals

埃及象形文的数字

14

15 Egyptian Hieroglyphic Numerals

埃及象形文的数字

15

16 Egyptian Hieroglyphic Numerals

埃及象形文的数字

16

17 Egyptian Hieroglyphic Numerals

埃及象形文的数字

17

18 Egyptian Hieroglyphic Numerals
埃及象形文的数字

18

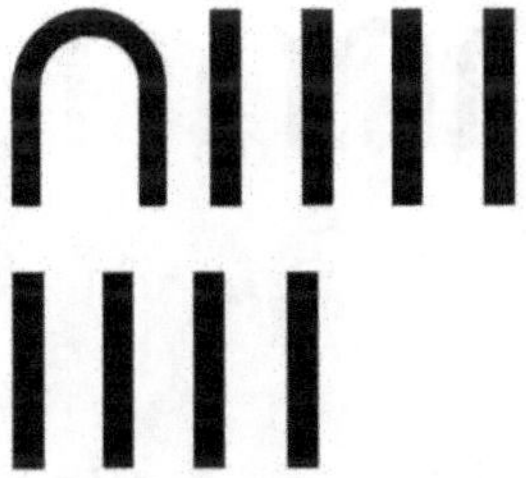

19 Egyptian Hieroglyphic Numerals
埃及象形文的数字

19

20 Egyptian Hieroglyphic Numerals
埃及象形文的数字

ḏubāʻatay 20

30 Egyptian Hieroglyphic Numerals
埃及象形文的数字

máʻbAʼ (masc.)
máʻbAʼat (fem.) 30

40 Egyptian Hieroglyphic Numerals

埃及象形文的数字

ḥAmí (masc.) 40

𓎆𓎆𓎆𓎆

50 Egyptian Hieroglyphic Numerals

埃及象形文的数字

díywu 50

𓎆𓎆𓎆𓎆𓎆

60 Egyptian Hieroglyphic Numerals

埃及象形文的数字

yAssáwyu 60

𓎆𓎆𓎆
𓎆𓎆𓎆

70 Egyptian Hieroglyphic Numerals

埃及象形文的数字

safḫáwyu (masc.) 70

𓎆𓎆𓎆𓎆
𓎆𓎆𓎆

80 Egyptian Hieroglyphic Numerals

埃及象形文的数字

ḫamanáwyu (masc.) 80

𓎆𓎆𓎆𓎆
𓎆𓎆𓎆𓎆

90 Egyptian Hieroglyphic Numerals

埃及象形文的数字

pAsiḏawyu (masc.) 90

𓎆𓎆𓎆𓎆𓎆
𓎆𓎆𓎆𓎆

100 Egyptian Hieroglyphic Numerals

埃及象形文的数字

šáwat 100

𓍢

200 Egyptian Hieroglyphic Numerals

埃及象形文的数字

šūtay 200

𓍢𓍢

300 Egyptian Hieroglyphic Numerals
埃及象形文的数字

300

400 Egyptian Hieroglyphic Numerals
埃及象形文的数字

400

500 Egyptian Hieroglyphic Numerals
埃及象形文的数字

500

600 Egyptian Hieroglyphic Numerals
埃及象形文的数字

600

700 Egyptian Hieroglyphic Numerals
埃及象形文的数字

700

800 Egyptian Hieroglyphic Numerals
埃及象形文的数字

800

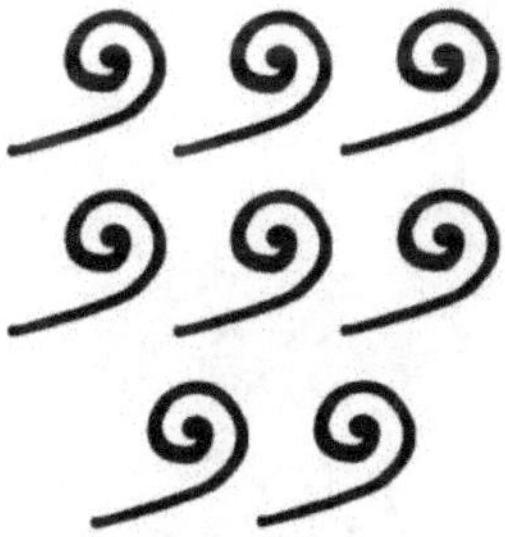

900 Egyptian Hieroglyphic Numerals
埃及象形文的数字

900

1,000 Egyptian Hieroglyphic Numerals
埃及象形文的数字

ḥa' 1,000

2,000 Egyptian Hieroglyphic Numerals
埃及象形文的数字

2,000

3,000 Egyptian Hieroglyphic Numerals
埃及象形文的数字

3,000

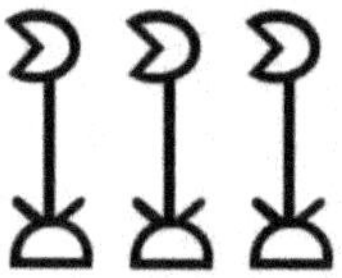

4,000 Egyptian Hieroglyphic Numerals
埃及象形文的数字

4,000

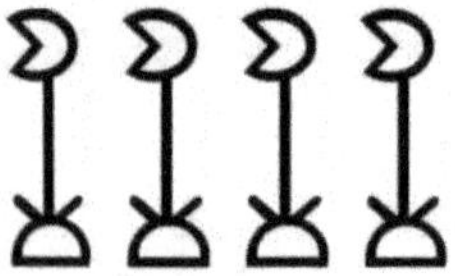

4,622 Egyptian Hieroglyphic Numerals
埃及象形文的数字

4,622

5,000 Egyptian Hieroglyphic Numerals
埃及象形文的数字

5,000

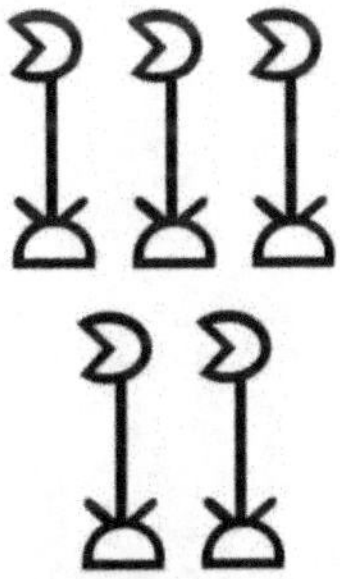

6,000 Egyptian Hieroglyphic Numerals
埃及象形文的数字

6,000

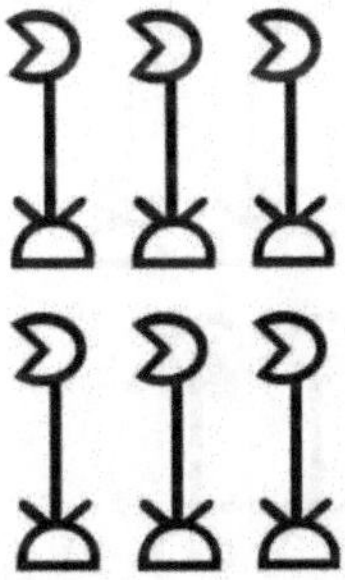

7,000 Egyptian Hieroglyphic Numerals
埃及象形文的数字

7,000

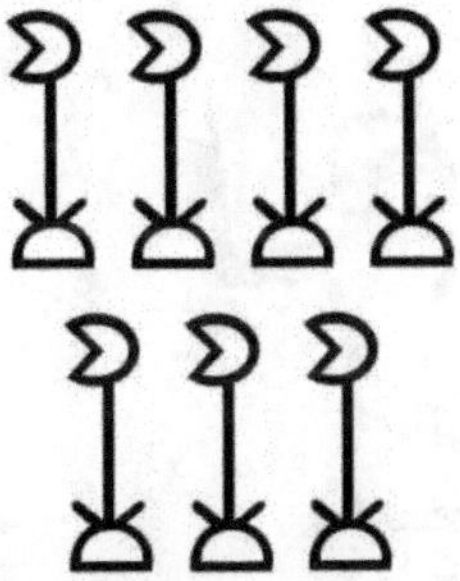

8,000 Egyptian Hieroglyphic Numerals
埃及象形文的数字

8,000

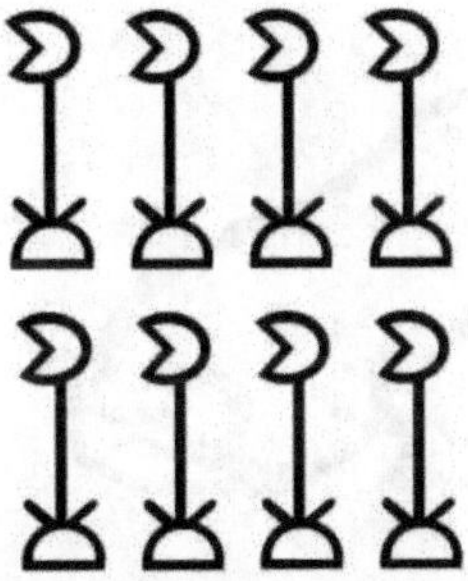

9,000 Egyptian Hieroglyphic Numerals

埃及象形文的数字

9,000

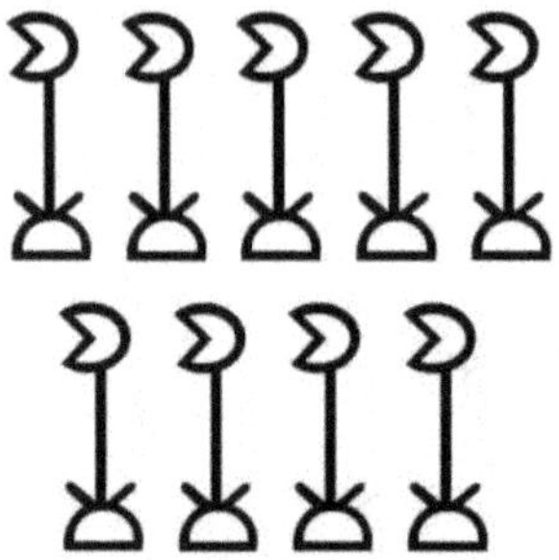

10,000 Egyptian Hieroglyphic Numerals

埃及象形文的数字

ḏuba‘ 10,000

100,000 Egyptian Hieroglyphic Numerals

埃及象形文的数字

hfn 100,000

1,000,000 Egyptian Hieroglyphic Numerals

埃及象形文的数字

ḥaḥ 1,000,000

Egyptian Hieratic Numerals 古埃及僧侣体数字

Hieratic numerals were written on papyrus and *ostraca* (broken pieces of pottery). The hieratic script was often used for writing ephemera, such as notes, medical prescriptions, accounting records and receipts and it contrasts with the more serious nature of hieroglyphic texts which were carved on stone. Broken pottery and papyrus were also a cheap and plentiful medium on which to practice writing. Hieratic numerals started to appear during the Early Dynastic Period (3150-2686 BC). Unlike hieroglyphic numerals, they could not be repeated to express the desired value. There were separate hieratic symbols for 1 to 9 as well as the tens, hundreds and thousands. By 650 BC Demotic script which was written right to left started to replace hieratic. Demotic declined in the third century BC when Greek became more important.

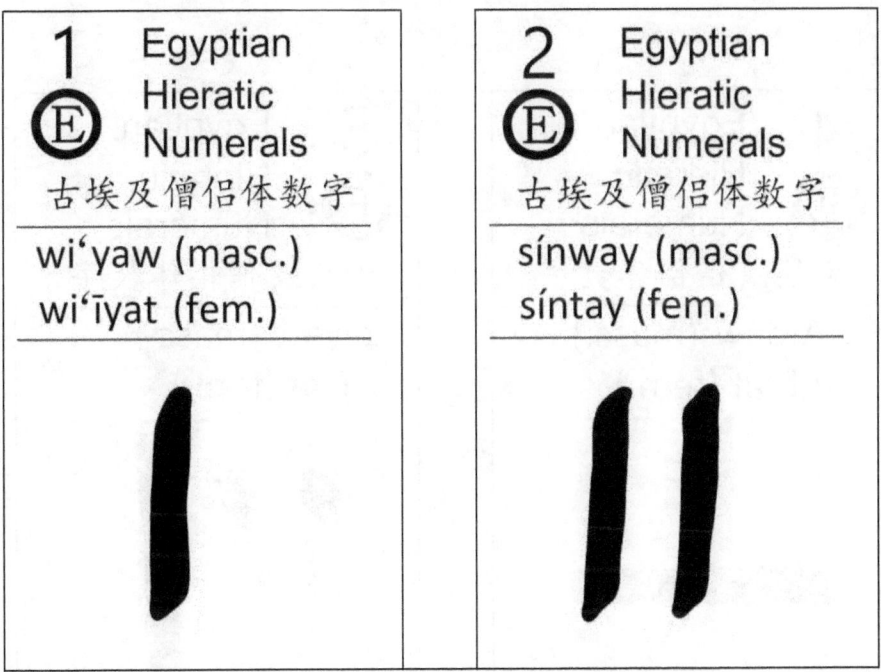

1 Ⓔ Egyptian Hieratic Numerals
古埃及僧侣体数字

wiʿyaw (masc.)
wiʿīyat (fem.)

2 Ⓔ Egyptian Hieratic Numerals
古埃及僧侣体数字

sínway (masc.)
síntay (fem.)

3 Ⓔ Egyptian Hieratic Numerals

古埃及僧侣体数字

ḫámtaw (masc.)
ḫámtat (fem.)

4 Ⓔ Egyptian Hieratic Numerals

古埃及僧侣体数字

yAfdáw (masc.)
yAfdát (fem.)

4 Ⓔ Egyptian Hieratic Numerals

古埃及僧侣体数字

yAfdáw (masc.)
yAfdát (fem.)

5 Ⓔ Egyptian Hieratic Numerals

古埃及僧侣体数字

dīyaw (masc.)
dīyat (fem.)

6 Ⓔ Egyptian Hieratic Numerals
古埃及僧侣体数字

yAssáw (masc.)
yAssát (fem.)

6 Ⓔ Egyptian Hieratic Numerals
古埃及僧侣体数字

yAssáw (masc.)
yAssát (fem.)

7 Ⓔ Egyptian Hieratic Numerals
古埃及僧侣体数字

sáfḫaw (masc.)
sáfḫat (fem.)

8 Ⓔ Egyptian Hieratic Numerals
古埃及僧侣体数字

ḫAmānaw (masc.)
ḫAmānat (fem.)

9 Egyptian Hieratic Numerals

古埃及僧侣体数字

pAsīḏaw (masc.)

pAsīḏat (fem.)

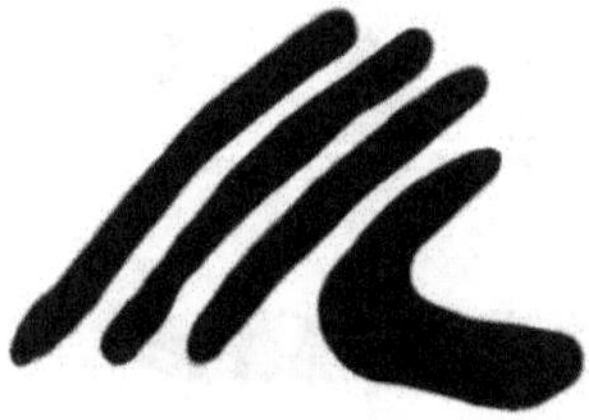

10 Egyptian Hieratic Numerals

古埃及僧侣体数字

mūḏaw (masc.)

mūḏat (fem.)

11 Egyptian Hieratic Numerals

古埃及僧侣体数字

11

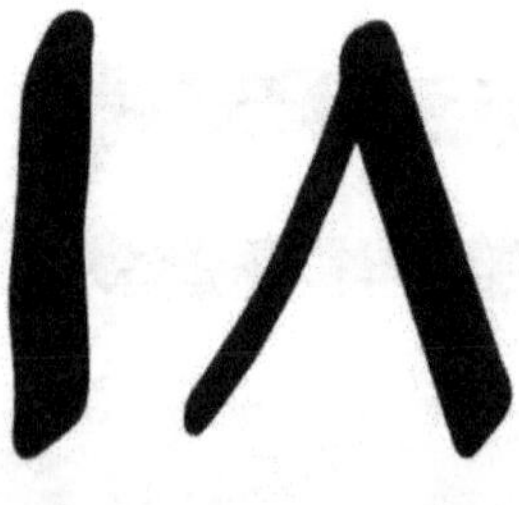

12 Egyptian Hieratic Numerals

古埃及僧侣体数字

12

13 Egyptian Hieratic Numerals
古埃及僧侣体数字

13

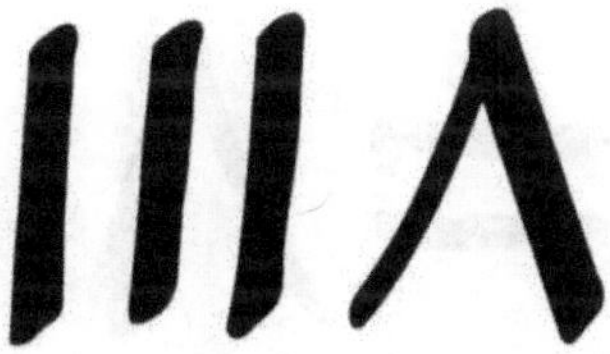

14 Egyptian Hieratic Numerals
古埃及僧侣体数字

14

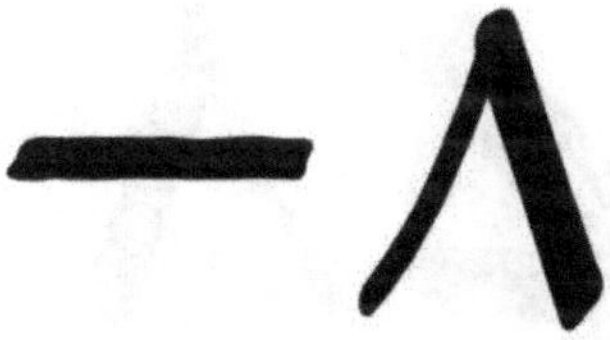

15 Egyptian Hieratic Numerals
古埃及僧侣体数字

15

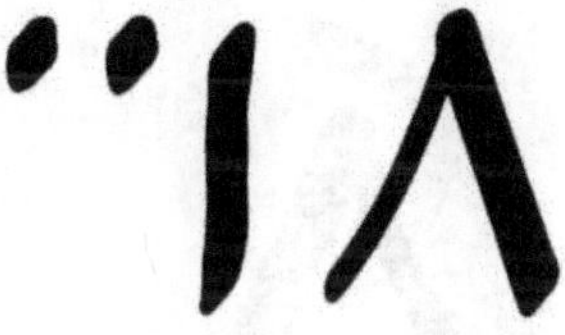

16 Egyptian Hieratic Numerals
古埃及僧侣体数字

16

17 Egyptian Hieratic Numerals
Ⓔ
古埃及僧侣体数字

17

18 Egyptian Hieratic Numerals
Ⓔ
古埃及僧侣体数字

18

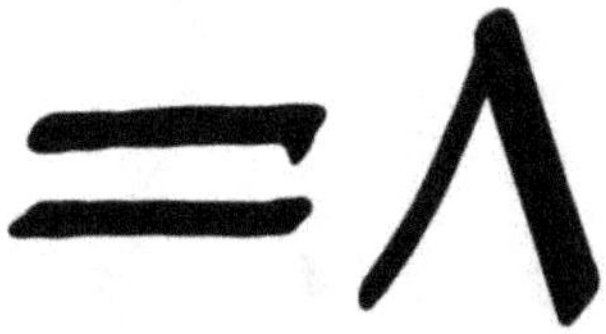

19 Egyptian Hieratic Numerals
Ⓔ
古埃及僧侣体数字

19

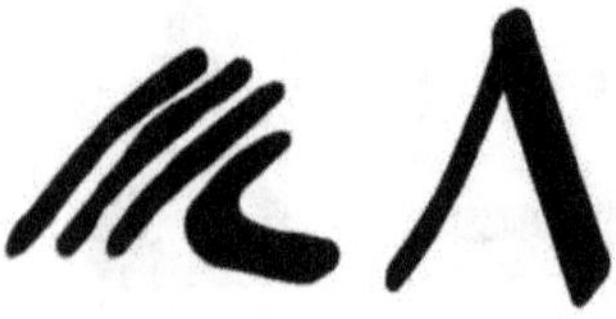

20 Egyptian Hieratic Numerals
Ⓔ
古埃及僧侣体数字

ḏubāʻatay 20

30 Egyptian Hieratic Numerals
Ⓔ
古埃及僧侣体数字

má'bA' (masc.) 30
má'bA'at (fem.)

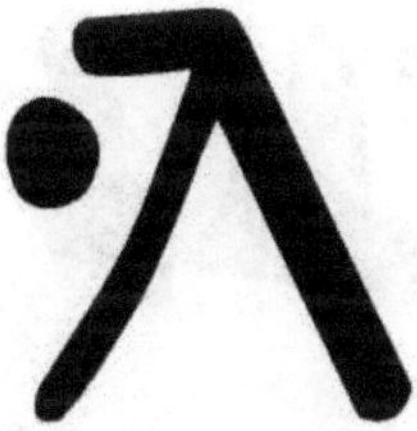

40 Egyptian Hieratic Numerals
Ⓔ
古埃及僧侣体数字

ḥAmí (masc.) 40

50 Egyptian Hieratic Numerals
Ⓔ
古埃及僧侣体数字

díywu 50

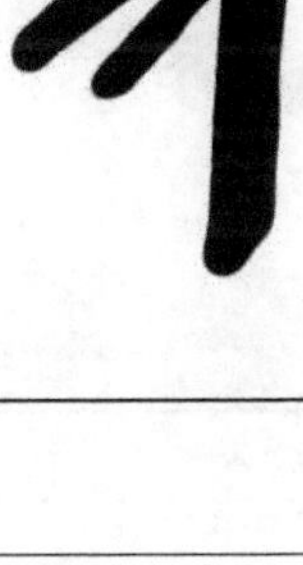

60 Egyptian Hieratic Numerals
Ⓔ
古埃及僧侣体数字

yAssáwyu 60

70 Ⓔ Egyptian Hieratic Numerals
古埃及僧侣体数字

safḫáwyu (masc.) 70

80 Ⓔ Egyptian Hieratic Numerals
古埃及僧侣体数字

ḫamanáwyu (masc.) 80

90 Ⓔ Egyptian Hieratic Numerals
古埃及僧侣体数字

pAsiḏawyu (masc.) 90

100 Ⓔ Egyptian Hieratic Numerals
古埃及僧侣体数字

šáwat 100

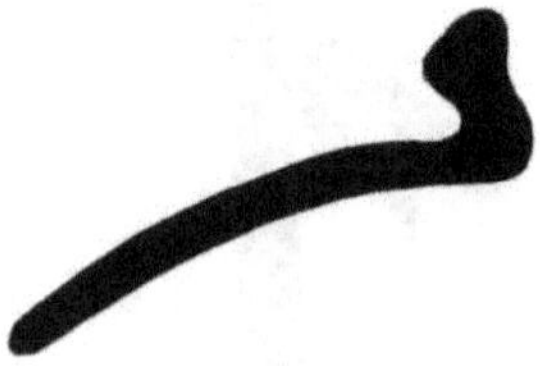

200 Egyptian Hieratic Numerals
Ⓔ
古埃及僧侣体数字

šūtay 200

300 Egyptian Hieratic Numerals
Ⓔ
古埃及僧侣体数字

300

400 Egyptian Hieratic Numerals
Ⓔ
古埃及僧侣体数字

400

500 Egyptian Hieratic Numerals
Ⓔ
古埃及僧侣体数字

500

600 Egyptian Hieratic Numerals Ⓔ
古埃及僧侣体数字

600

700 Egyptian Hieratic Numerals Ⓔ
古埃及僧侣体数字

700

800 Egyptian Hieratic Numerals Ⓔ
古埃及僧侣体数字

800

900 Egyptian Hieratic Numerals Ⓔ
古埃及僧侣体数字

900

1,000 Egyptian Hieratic Numerals Ⓔ

古埃及僧侣体数字

ḫa’ 1,000

1,000 Egyptian Hieratic Numerals Ⓔ

古埃及僧侣体数字

ḫa’ 1,000

2,000 Egyptian Hieratic Numerals Ⓔ

古埃及僧侣体数字

2,000

3,000 Egyptian Hieratic Numerals Ⓔ

古埃及僧侣体数字

3,000

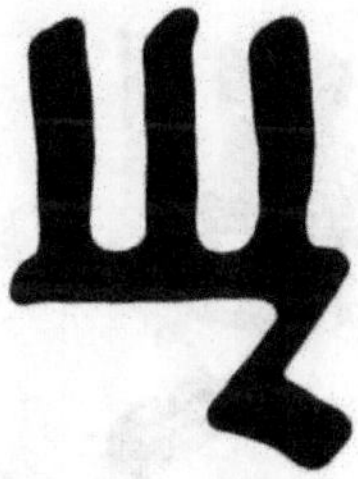

4,000 Egyptian Hieratic Numerals
Ⓔ
古埃及僧侣体数字

4,000

5,000 Egyptian Hieratic Numerals
Ⓔ
古埃及僧侣体数字

5,000

6,000 Egyptian Hieratic Numerals
Ⓔ
古埃及僧侣体数字

6,000

7,000 Egyptian Hieratic Numerals
Ⓔ
古埃及僧侣体数字

7,000

8,000 Egyptian Hieratic Numerals

Ⓔ

古埃及僧侣体数字

8,000

9,000 Egyptian Hieratic Numerals

Ⓔ

古埃及僧侣体数字

9,000

10,000 Egyptian Hieratic Numerals

Ⓔ

古埃及僧侣体数字

ḏuba‘ 10,000

100,000 Egyptian Hieratic Numerals

Ⓔ

古埃及僧侣体数字

hfn 100,000

Aegean/Minoan Numerals 爱琴/ 米诺按数字

Aegean numerals appeared in the largely un-deciphered Linear A script found on the island of Crete and dated to the first half of the 2nd millennium BC. However, by 1400 BC they were used in the Linear B script as a fully developed decimal system based on additive principles and perhaps inspired by Egyptian hieroglyphic numerals. A single vertical stroke represents one which is repeated to form a group of nine strokes symbolizing the number nine, just as in Egyptian hieroglyphs. Similarly a circle represents 100 while a group of nine circles indicates 900.

Linear B found in both Crete and the Peloponnese of mainland Greece was deciphered by Michael Ventris in 1952 and shown to be an archaic form of Greek. The un-deciphered Cypro-Minoan syllabary used on the island of Cyprus from about 1550 to 1050 BC is similar to Linear A in appearance and may have been derived from it. The Cypriot syllabary (11th to 4th century BC), used to write the Arcadocypriot dialect of Greek, is in turn descended from the Cypro-Minoan syllabary. Studies of DNA from the island of Crete indicate that the ancient inhabitants were largely of Greek origin. This then shows that the first European civilization was located in Crete and that the Aegean/Minoan numerals were the first European numbering system.

1 Ⓑ Aegean/ Minoan Numerals

爱琴/米诺按数字

2 Ⓑ Aegean/ Minoan Numerals

爱琴/米诺按数字

3 Ⓑ Aegean/ Minoan Numerals

爱琴/米诺按数字

4 Ⓑ Aegean/ Minoan Numerals

爱琴/米诺按数字

5

β

Aegean/ Minoan Numerals

爱琴/米诺按数字

6

β

Aegean/ Minoan Numerals

爱琴/米诺按数字

7

β

Aegean/ Minoan Numerals

爱琴/米诺按数字

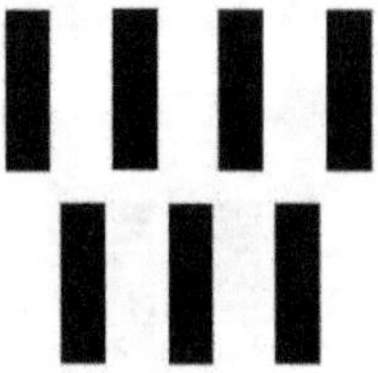

8

β

Aegean/ Minoan Numerals

爱琴/米诺按数字

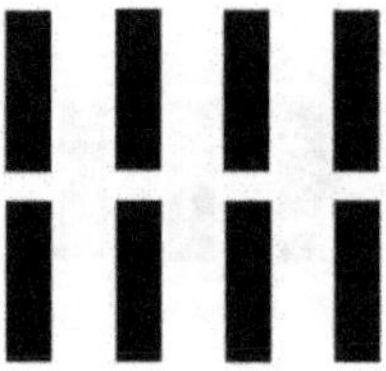

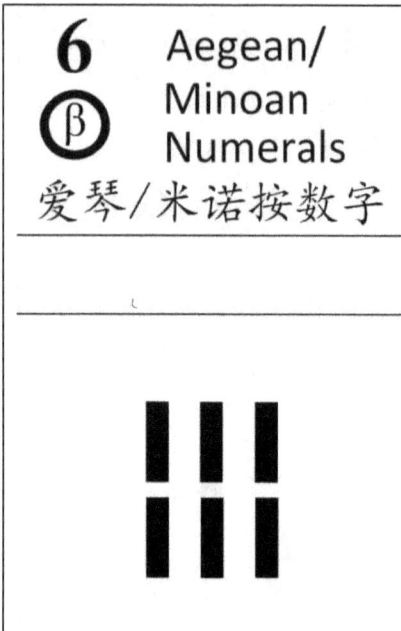

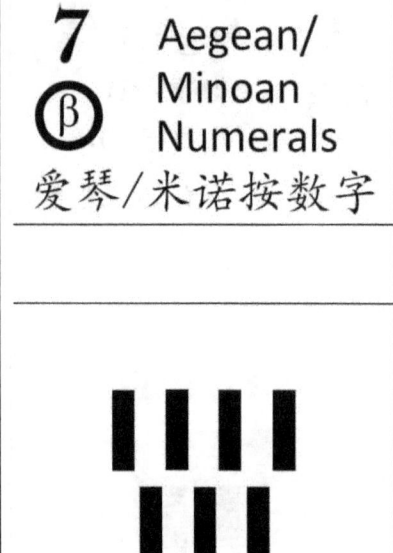

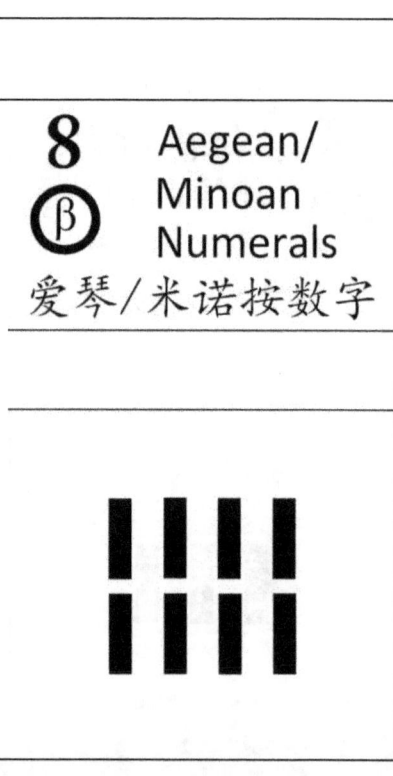

9 Aegean/ Minoan Numerals
ⓑ
爱琴/米诺按数字

𐄏

10 Aegean/ Minoan Numerals
ⓑ
爱琴/米诺按数字

𐄐

11 Aegean/ Minoan Numerals
ⓑ
爱琴/米诺按数字

𐄐𐄇

12 Aegean/ Minoan Numerals
ⓑ
爱琴/米诺按数字

𐄐𐄈

13 (β) Aegean/ Minoan Numerals

爱琴/米诺按数字

14 (β) Aegean/ Minoan Numerals

爱琴/米诺按数字

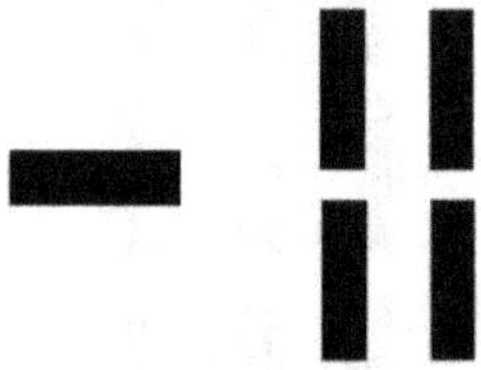

15 (β) Aegean/ Minoan Numerals

爱琴/米诺按数字

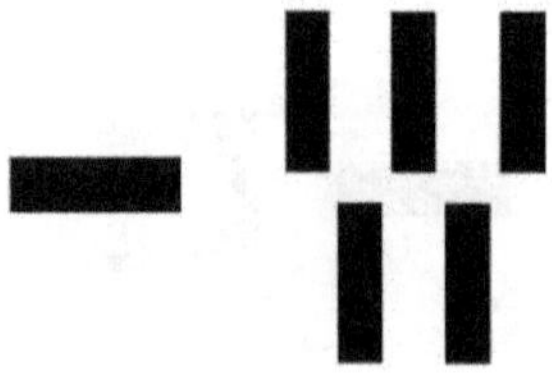

16 (β) Aegean/ Minoan Numerals

爱琴/米诺按数字

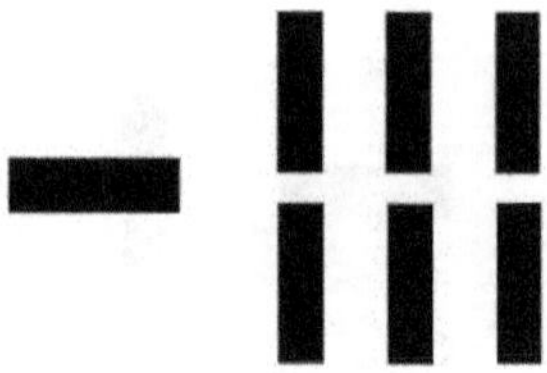

17 Aegean/ Minoan Numerals
ⓑ
爱琴/米诺按数字

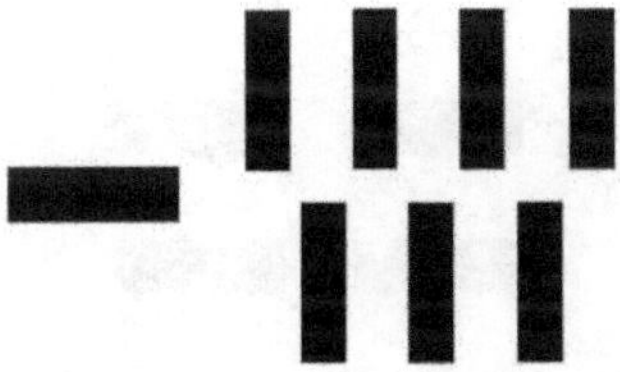

18 Aegean/ Minoan Numerals
ⓑ
爱琴/米诺按数字

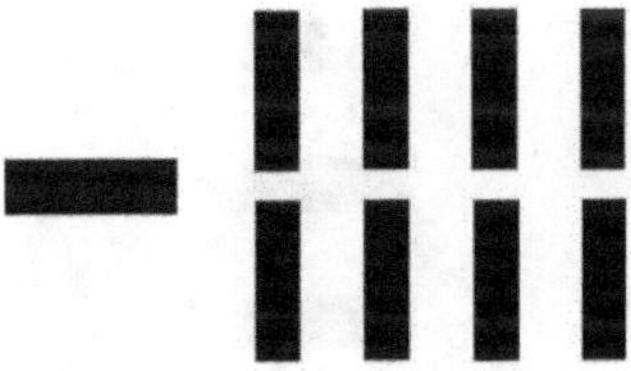

19 Aegean/ Minoan Numerals
ⓑ
爱琴/米诺按数字

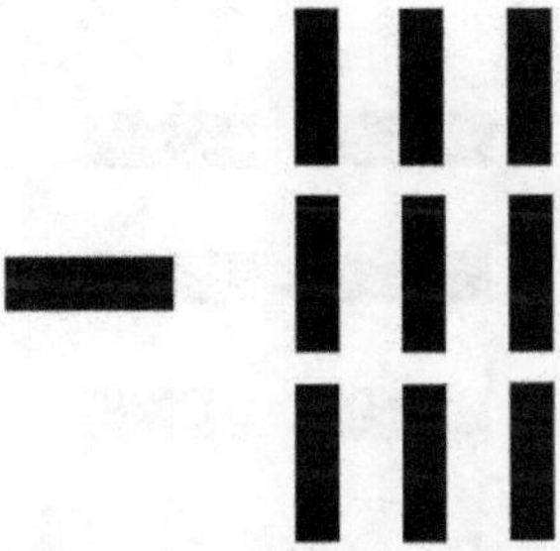

20 Aegean/ Minoan Numerals
ⓑ
爱琴/米诺按数字

30 (β) Aegean/ Minoan Numerals

爱琴/米诺按数字

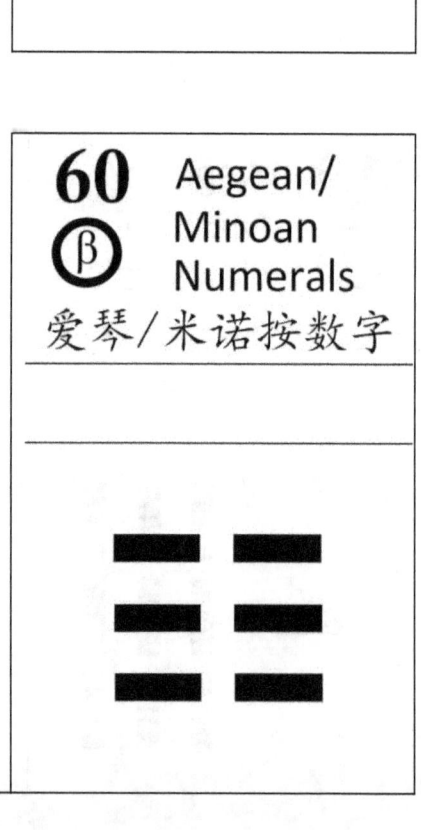

40 (β) Aegean/ Minoan Numerals

爱琴/米诺按数字

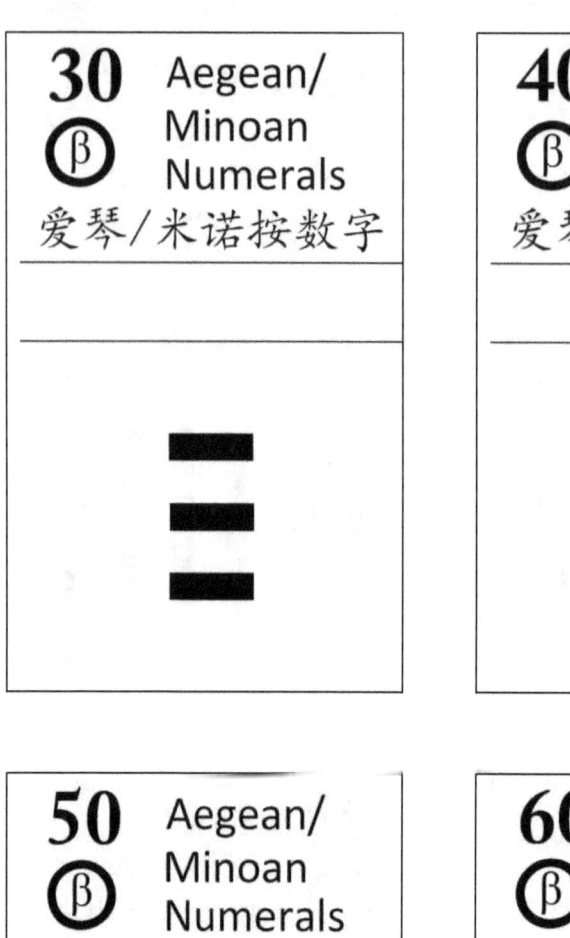

50 (β) Aegean/ Minoan Numerals

爱琴/米诺按数字

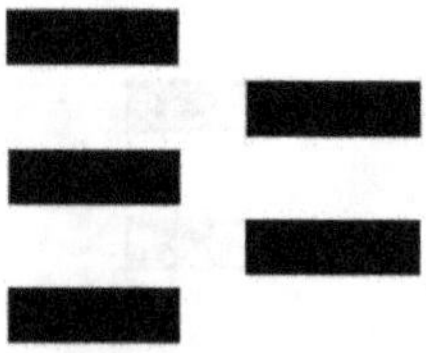

60 (β) Aegean/ Minoan Numerals

爱琴/米诺按数字

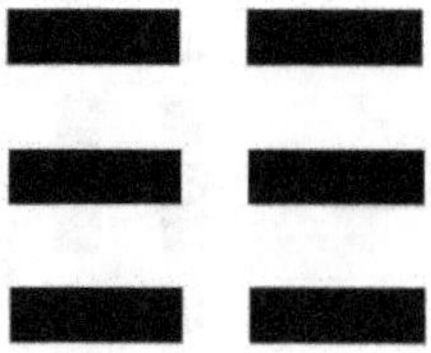

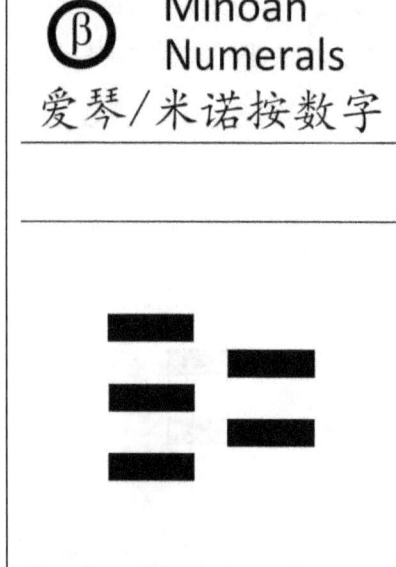

70 Aegean/ Minoan Numerals

(β)

爱琴/米诺按数字

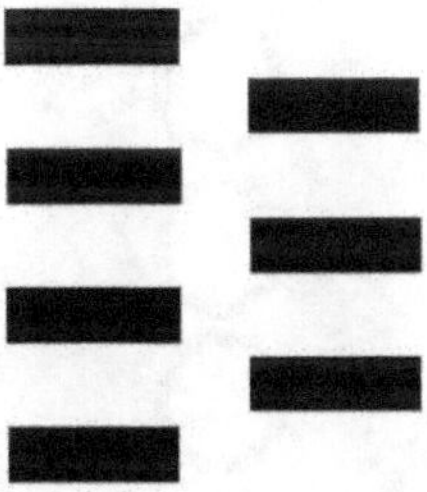

80 Aegean/ Minoan Numerals

(β)

爱琴/米诺按数字

90 Aegean/ Minoan Numerals

(β)

爱琴/米诺按数字

100 Aegean/ Minoan Numerals

(β)

爱琴/米诺按数字

200 (β) Aegean/ Minoan Numerals

爱琴/米诺按数字

300 (β) Aegean/ Minoan Numerals

爱琴/米诺按数字

400 (β) Aegean/ Minoan Numerals

爱琴/米诺按数字

500 (β) Aegean/ Minoan Numerals

爱琴/米诺按数字

600 Aegean/ Minoan Numerals
β
爱琴/米诺按数字

700 Aegean/ Minoan Numerals
β
爱琴/米诺按数字

800 Aegean/ Minoan Numerals
β
爱琴/米诺按数字

900 Aegean/ Minoan Numerals
β
爱琴/米诺按数字

1,000 Aegean/ Minoan Numerals

β

爱琴/米诺按数字

2,000 Aegean/ Minoan Numerals

β

爱琴/米诺按数字

3,000 Aegean/ Minoan Numerals

β

爱琴/米诺按数字

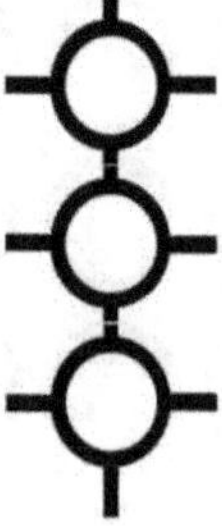

4,000 Aegean/ Minoan Numerals

β

爱琴/米诺按数字

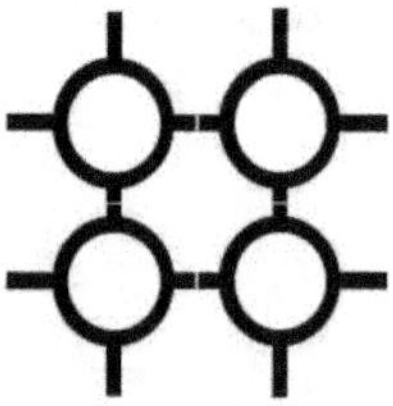

5,000 Aegean/
Minoan
Numerals
爱琴/米诺按数字

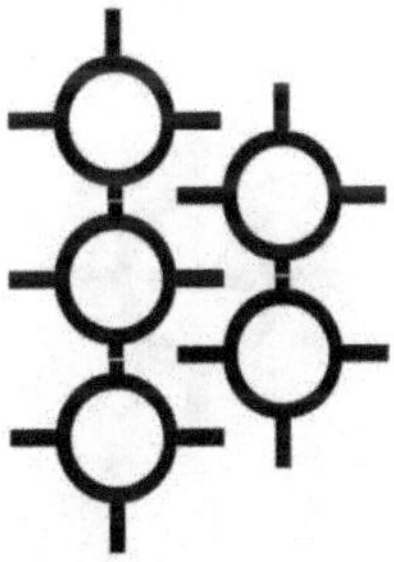

6,000 Aegean/
Minoan
Numerals
爱琴/米诺按数字

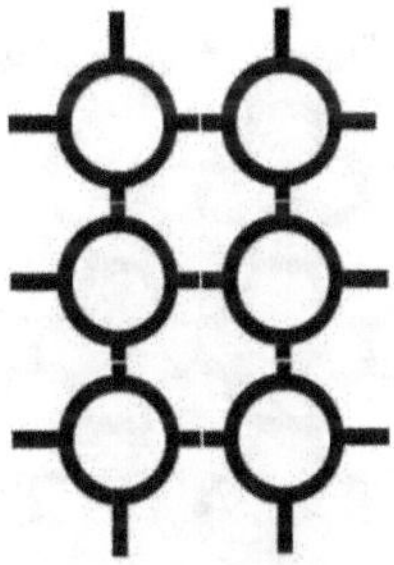

7,000 Aegean/
Minoan
Numerals
爱琴/米诺按数字

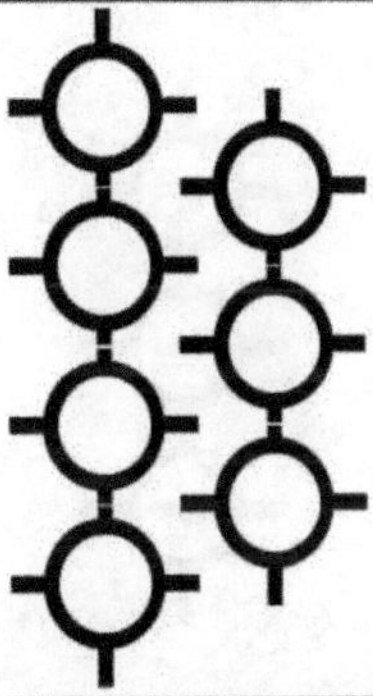

8,000 Aegean/
Minoan
Numerals
爱琴/米诺按数字

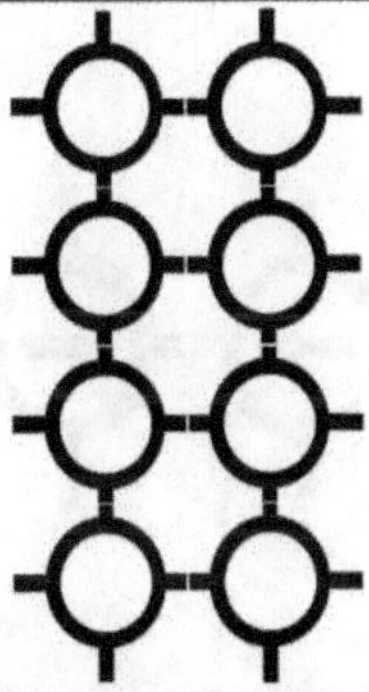

9,000 Aegean/ Minoan Numerals
β
爱琴/米诺按数字

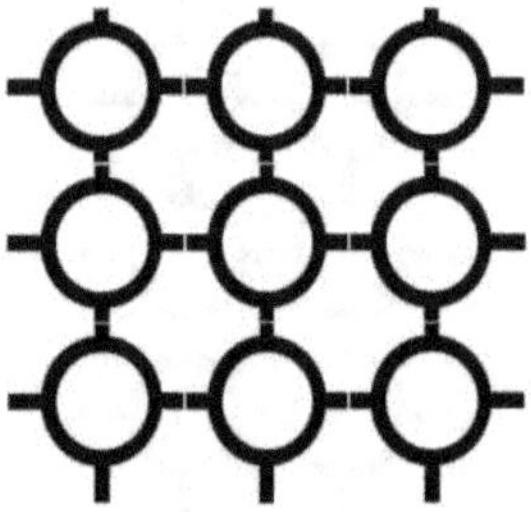

10,000 Aegean/ Minoan Numerals
β
爱琴/米诺按数字

20,000 Aegean/ Minoan Numerals
β
爱琴/米诺按数字

30,000 Aegean/ Minoan Numerals
β
爱琴/米诺按数字

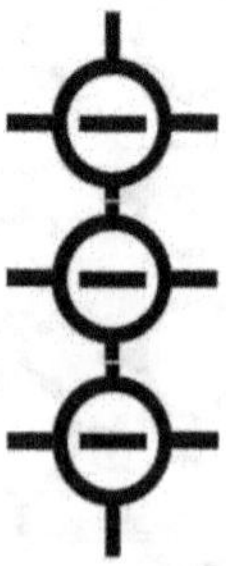

40,000 Aegean/ Minoan Numerals

爱琴/米诺按数字

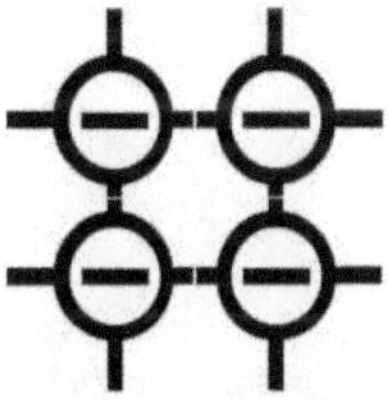

50,000 Aegean/ Minoan Numerals

爱琴/米诺按数字

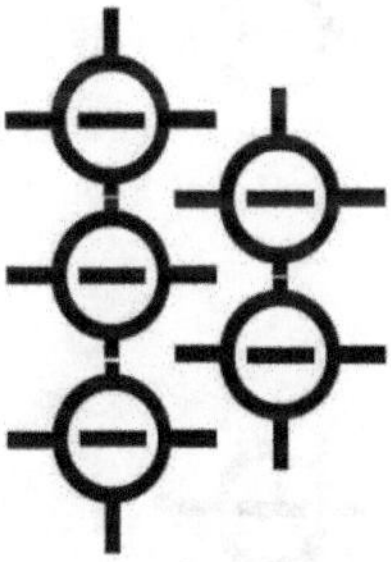

52,964 Aegean/ Minoan Numerals

爱琴/米诺按数字

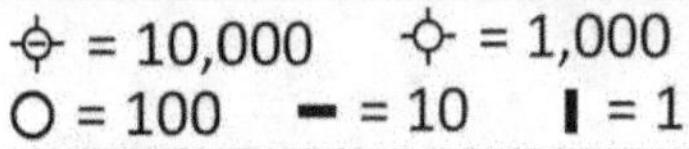

52,964

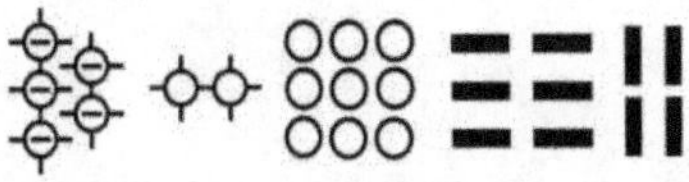

99,999

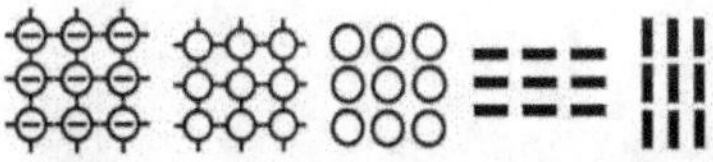

60,000 Aegean/ Minoan Numerals

爱琴/米诺按数字

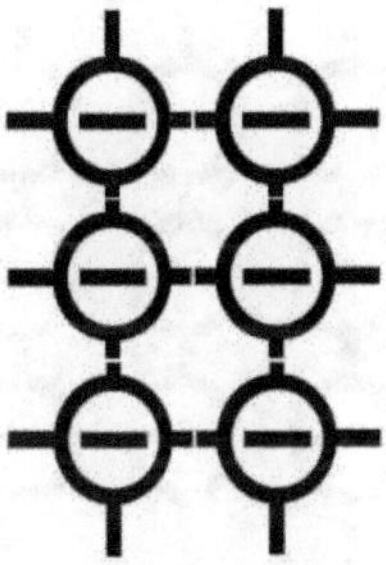

70,000 Aegean/ Minoan Numerals

爱琴/米诺按数字

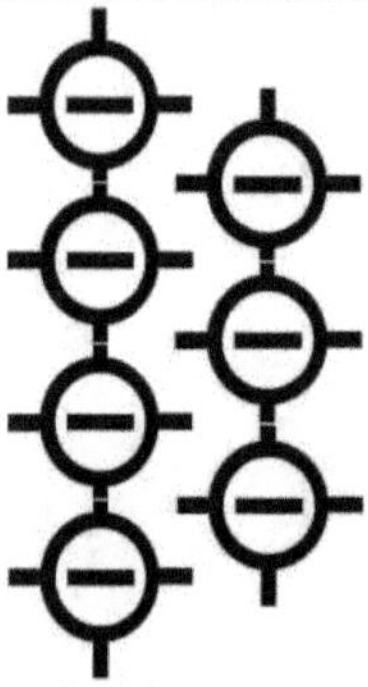

80,000 Aegean/ Minoan Numerals

爱琴/米诺按数字

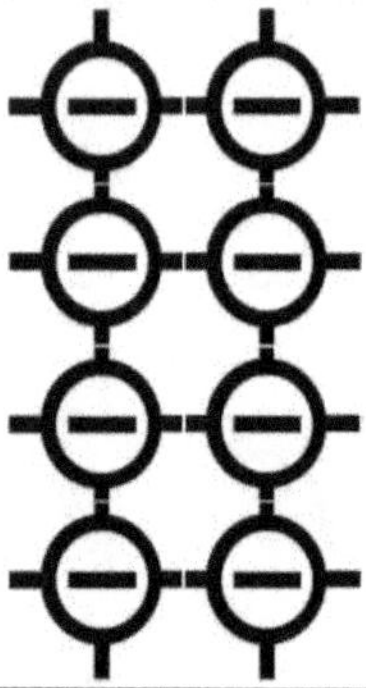

90,000 Aegean/ Minoan Numerals

爱琴/米诺按数字

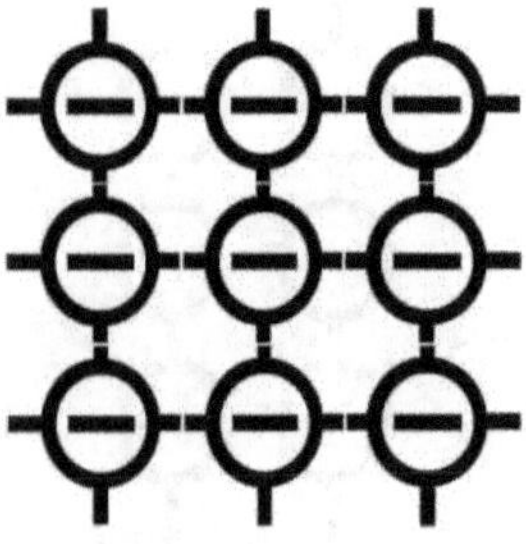

Phoenician Numerals 腓尼基数字

The Phoenician numeral system was written from right to left and had separate symbols for one, ten, twenty, hundred and thousand but there was no symbol for zero. Symbols could be repeated many times to represent the desired value and this contrasted adversely with the Egyptian hieratic numerals which did not allow repetition. One was a simple vertical stroke and such strokes were arranged in groups of three for numbers up to nine. Another problem was that there were at least six variants of the symbol for 20 and several variants of the symbol for hundred. The digits one, two or three etc. could precede the symbol for hundred to indicate the numbers 100 up to 900.

It is important to show the letters of the Phoenician alphabet which were copied and adapted by the Greeks and eventually used as numerals, not only by the Greeks but eventually by Hebrews, Arabs Ethiopians and others. The Phoenician alphabet has its origins in inscriptions in the Proto-Sinaitic script, also called Proto-Canaanite, the earliest examples of which are dated between the mid 19th and mid 16th centuries BC. These inscriptions, which were often crudely done, were influenced by Egyptian hieroglyphic writing. Hieroglyphs may have been radically simplified and then translated into the Semitic language of Canaan and Lebanon. The translated word and its graphic simplification may then have been used as a letter in the newly invented alphabet. After centuries of development, the earliest extant inscription in the Phoenician alphabet is dated 1050 BC. Here are the letters and their names and the objects and animals which may have provided the inspiration for the alphabet.

𐤀	𐤁	𐤂	𐤃	𐤄	𐤅	𐤆
Aleph Ox	Beth House	Gimel Camel	Daleth Door	He Window	Waw Hook	Zayin Weapon

𐤇	𐤈	𐤉	𐤊	𐤋	𐤌	𐤍	𐤎
Heth Fence, wall	Teth Wheel	Yodh Hand	Kaph Palm of hand	Lamedh Ox goad	Mem Water	Nun Snake	Samekh Fish

𐤏	𐤐	𐤑	𐤒	𐤓	𐤔	𐤕
Ayin Eye	Peh Mouth	Sadhe Fishhook	Qoph Monkey	Resh Head	Shin Tooth	Taw Mark

1 Phoenician Numerals 腓尼基数字

Phoenician Script Right to Left: 𐤀𐤇𐤃

Transcription Left to Right Vowels Unknown: ʼḥd

𐤖

2 Phoenician Numerals 腓尼基数字

Phoenician Script Right to Left: 𐤔𐤍𐤌

Transcription Left to Right Vowels Unknown: šnm

𐤚

3 Ⓟ Phoenician Numerals 腓尼基数字

Phoenician Script Right to Left: 𐤔𐤋𐤔

Transcription Left to Right Vowels Unknown: šlš

4 Ⓟ Phoenician Numerals 腓尼基数字

Phoenician Script Right to Left: 𐤀𐤓𐤁

Transcription Left to Right Vowels Unknown: ’rb

5 Ⓟ Phoenician Numerals 腓尼基数字

Phoenician Script Right to Left: 𐤇𐤌𐤔

Transcription Left to Right Vowels Unknown: ḥmš

6 Ⓟ Phoenician Numerals 腓尼基数字

Phoenician Script Right to Left: 𐤔𐤔

Transcription Left to Right Vowels Unknown: šš

7 Ⓟ Phoenician Numerals 腓尼基数字

Phoenician Script
Right to Left — 𐤔𐤁𐤏

Transcription
Left to Right
Vowels Unknown — šbʿ

8 Ⓟ Phoenician Numerals 腓尼基数字

Phoenician Script
Right to Left — 𐤔𐤌𐤍

Transcription
Left to Right
Vowels Unknown — šmn

9 Ⓟ Phoenician Numerals 腓尼基数字

Phoenician Script
Right to Left — 𐤕𐤔𐤏

Transcription
Left to Right
Vowels Unknown — tšʿ

10 Ⓟ Phoenician Numerals 腓尼基数字

Phoenician Script
Right to Left — 𐤏𐤔𐤓

Transcription
Left to Right
Vowels Unknown — ʿšr

11 Ⓟ Phoenician Numerals 腓尼基数字

𐤀𐤇𐤃 𐤏𐤔𐤓

ʼḥd ʻšr

12 Ⓟ Phoenician Numerals 腓尼基数字

𐤔𐤍𐤌 𐤏𐤔𐤓

šnm ʻšr

13 Ⓟ Phoenician Numerals 腓尼基数字

𐤔𐤋𐤔 𐤏𐤔𐤓

šlš ʻšr

14 Ⓟ Phoenician Numerals 腓尼基数字

𐤀𐤓𐤁 𐤏𐤔𐤓

ʼrb ʻšr

15 Phoenician Numerals 腓尼基数字

Ⓟ

𐤇𐤌𐤔 𐤏𐤔𐤓

ḥmš ʿšr

𐤗𐤛𐤚

16 Phoenician Numerals 腓尼基数字

Ⓟ

𐤔𐤔 𐤏𐤔𐤓

šš ʿšr

𐤗𐤛𐤛

17 Phoenician Numerals 腓尼基数字

Ⓟ

𐤔𐤁𐤏 𐤏𐤔𐤓

šbʿ ʿšr

𐤗𐤛𐤛𐤖

18 Phoenician Numerals 腓尼基数字

Ⓟ

𐤔𐤌𐤍 𐤏𐤔𐤓

šmn ʿšr

𐤗𐤛𐤛𐤚

19 Phoenician Numerals

Ⓟ 腓尼基数字

𐤓WO OWX

tšʻ ʻšr

20 Phoenician Numerals

Ⓟ 腓尼基数字

Several glyph variants for 20 display 2 parallel lines connected by a 3rd line.

20 Phoenician Numerals

Ⓟ 腓尼基数字

Several glyph variants for 20 display 2 parallel lines connected by a 3rd line.

20 Phoenician Numerals

Ⓟ 腓尼基数字

Several glyph variants for 20 display 2 parallel lines connected by a 3rd line.

20 Phoenician Numerals
Ⓟ 腓尼基数字

Several glyph variants for 20 display 2 parallel lines connected by a 3rd line.

20 Phoenician Numerals
Ⓟ 腓尼基数字

Several glyph variants for 20 display 2 parallel lines connected by a 3rd line.

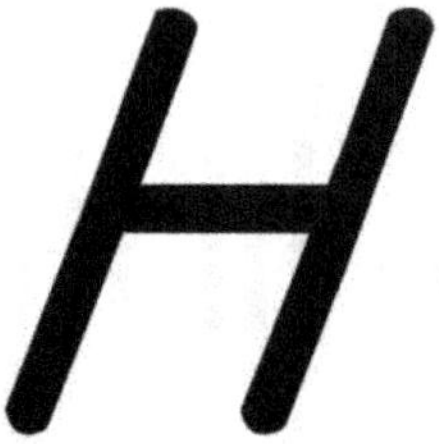

20 Phoenician Numerals
Ⓟ 腓尼基数字

Several glyph variants for 20 display 2 parallel lines connected by a 3rd line.

30 Phoenician Numerals
Ⓟ 腓尼基数字

40 Phoenician Numerals

Ⓟ 腓尼基数字

50 Phoenician Numerals

Ⓟ 腓尼基数字

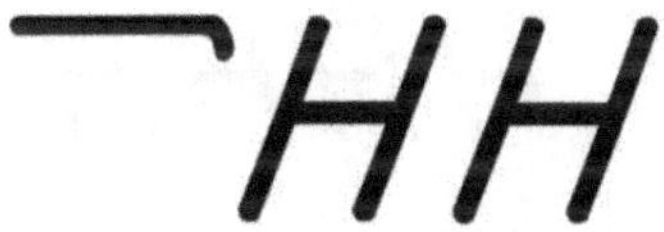

60 Phoenician Numerals

Ⓟ 腓尼基数字

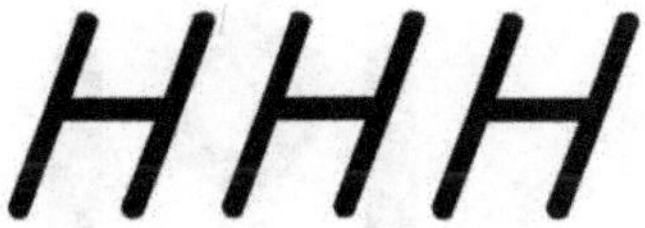

70 Phoenician Numerals

Ⓟ 腓尼基数字

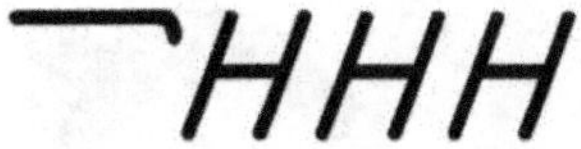

80 Phoenician Numerals

Ⓟ 腓尼基数字

HHHH

90 Phoenician Numerals

Ⓟ 腓尼基数字

100 Phoenician Numerals

Ⓟ 腓尼基数字

There are several glyph variants for 100.

100 Phoenician Numerals

Ⓟ 腓尼基数字

There are several glyph variants for 100.

100 Phoenician Numerals

Ⓟ 腓尼基数字

There are several glyph variants for 100.

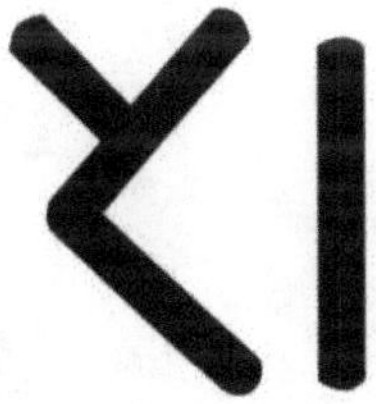

200 Phoenician Numerals

Ⓟ 腓尼基数字

200 Phoenician Numerals

Ⓟ 腓尼基数字

200 Phoenician Numerals

Ⓟ 腓尼基数字

300 Phoenician Numerals Ⓟ 腓尼基数字	**1,000** Phoenician Numerals Ⓟ 腓尼基数字

Attic Numerals 阿提卡数字

Attic numerals were used in ancient Attica and possibly date back to the 7th century BC although they were first described by Herodian in the second century. They are therefore also called Herodianic Numerals. Another name for them is acrophonic numerals because their symbols Π (πέντε - *pente*), Δ (δέκα - *deka*), H (ἑκατόν - *hekaton*), X (χίλιοι / χιλιάς – *khilioi/khilias*) and M (μύριον - *myrion*) were the first letters in the Greek words for five, ten, hundred, thousand and ten thousand. However, H was pronounced *eta* when the Ionic alphabet was later adopted. The symbol for five Π *pi* was written with a shorter right leg 𐅃 and symbols for 50, 500, 5,000 and 50,000 were made from the letters Δ, H, X and M placed inside the letter *pi* in this fashion; 𐅄, 𐅅, 𐅆, 𐅇. These composite symbols could be easily understood as 5 x 10; 5 x 100 ; 5 x 1,000 and 5 x 10,000.

1 Attic Acrophonic Numerals
阿提卡数字

Greek Alphabetic A′
alpha 1

Roman Numerals I

Acrophonic

I

2 Attic Acrophonic Numerals
阿提卡数字

Greek Alphabetic B′
beta 2

Roman Numerals II

Acrophonic

II

3 Attic Acrophonic Numerals
阿提卡数字

Greek Alphabetic Γ′
gamma 3

Roman Numerals III

Acrophonic

III

4 Attic Acrophonic Numerals
阿提卡数字

Greek Alphabetic Δ′
delta 4

Roman Numerals IV

Acrophonic

IIII

5 Attic Acrophonic Numerals 阿提卡数字

Greek Alphabetic E′
epsilon 5

Roman Numerals V

Acrophonic

𐅃

6 Attic Acrophonic Numerals 阿提卡数字

Greek Alphabetic F′
digamma 6

Roman Numerals VI

Acrophonic

𐅃I

7 Attic Acrophonic Numerals 阿提卡数字

Greek Alphabetic Z′
zeta 7

Roman Numerals VII

Acrophonic

𐅃II

8 Attic Acrophonic Numerals 阿提卡数字

Greek Alphabetic H′
eta 8

Roman Numerals VIII

Acrophonic

𐅃III

9 ⓐ Attic Acrophonic Numerals
阿提卡数字

Greek Alphabetic Θ′
theta 9

Roman Numerals IX

Acrophonic

𐅃IIII

10 ⓐ Attic Acrophonic Numerals
阿提卡数字

Greek Alphabetic Ι′
iota 10

Roman Numerals X

Acrophonic

Δ

11 ⓐ Attic Acrophonic Numerals
阿提卡数字

Greek Alphabetic ΙΑ′
11

Roman Numerals XI

Acrophonic

ΔI

12 ⓐ Attic Acrophonic Numerals
阿提卡数字

Greek Alphabetic ΙΒ′
12

Roman Numerals XII

Acrophonic

ΔII

13 @	Attic Acrophonic Numerals 阿提卡数字
Greek Alphabetic	IΓ′ 13
Roman Numerals	XIII
Acrophonic	ΔIII

14 @	Attic Acrophonic Numerals 阿提卡数字
Greek Alphabetic	IΔ′ 14
Roman Numerals	XIV
Acrophonic	ΔIIII

15 @	Attic Acrophonic Numerals 阿提卡数字
Greek Alphabetic	IE′ 15
Roman Numerals	XV
Acrophonic	ΔΓ

16 @	Attic Acrophonic Numerals 阿提卡数字
Greek Alphabetic	IϜ′ 16
Roman Numerals	XVI
Acrophonic	ΔΓI

17 Attic Acrophonic Numerals
阿提卡数字

Greek Alphabetic	IZ′ 17
Roman Numerals	XVII
Acrophonic	ΔΓII

18 Attic Acrophonic Numerals
阿提卡数字

Greek Alphabetic	IH′ 18
Roman Numerals	XVIII
Acrophonic	ΔΓIII

19 Attic Acrophonic Numerals
阿提卡数字

Greek Alphabetic	IΘ′ 19
Roman Numerals	XIX
Acrophonic	ΔΓIIII

20 Attic Acrophonic Numerals
阿提卡数字

Greek Alphabetic *kappa*	K′ 20
Roman Numerals	XX
Acrophonic	ΔΔ

30 Attic Acrophonic Numerals
阿提卡数字

Greek Alphabetic Λʹ
lambda 30

Roman Numerals XXX

Acrophonic

ΔΔΔ

40 Attic Acrophonic Numerals
阿提卡数字

Greek Alphabetic Μʹ
mu 40

Roman Numerals XL

Acrophonic

ΔΔΔΔ

50 Attic Acrophonic Numerals
阿提卡数字

Greek Alphabetic Νʹ
nu 50

Roman Numerals L

Acrophonic

𐅄

60 Attic Acrophonic Numerals
阿提卡数字

Greek Alphabetic Ξʹ
xi 60

Roman Numerals LX

Acrophonic

𐅄Δ

70 Attic Acrophonic Numerals

ⓐ 阿提卡数字

Greek Alphabetic O′

omicron 70

Roman Numerals LXX

Acrophonic

𐅄ΔΔ

80 Attic Acrophonic Numerals

ⓐ 阿提卡数字

Greek Alphabetic Π′

pi 80

Roman Numerals LXXX

Acrophonic

𐅄ΔΔΔ

90 Attic Acrophonic Numerals

ⓐ 阿提卡数字

Greek Alphabetic Ϙ′ ϟ′

koppa 90

Roman Numerals XC

Acrophonic

𐅄ΔΔΔΔ

100 Attic Acrophonic Numerals

ⓐ 阿提卡数字

Greek Alphabetic P′

rho 100

Roman Numerals C

Acrophonic

H

200 Attic Acrophonic Numerals

阿提卡数字

Greek Alphabetic Σʹ

sigma 200

Roman Numerals CC

Acrophonic

HH

300 Attic Acrophonic Numerals

阿提卡数字

Greek Alphabetic Τʹ

tau 300

Roman Numerals CCC

Acrophonic

HHH

400 Attic Acrophonic Numerals

阿提卡数字

Greek Alphabetic Υʹ

epsilon 400

Roman Numerals CD

Acrophonic

HHHH

500 Attic Acrophonic Numerals

阿提卡数字

Greek Alphabetic Φʹ

phi 500

Roman Numerals D

Acrophonic

𐅅

600 Attic Acrophonic Numerals 阿提卡数字

Greek Alphabetic *chi*	Χʹ 600
Roman Numerals	DC
Acrophonic	𐅅Η

700 Attic Acrophonic Numerals 阿提卡数字

Greek Alphabetic *psi*	Ψʹ 700
Roman Numerals	DCC
Acrophonic	𐅅ΗΗ

800 Attic Acrophonic Numerals 阿提卡数字

Greek Alphabetic *omega*	Ωʹ 800
Roman Numerals	DCCC
Acrophonic	𐅅ΗΗΗ

900 Attic Acrophonic Numerals 阿提卡数字

Greek Alphabetic *sampi*	ϡʹ 900
Roman Numerals	CM
Acrophonic	𐅅ΗΗΗΗ

1,000 Attic Acrophonic Numerals
阿提卡数字

Greek Alphabetic ͵Α′ 1,000
Subscript *iota* before *alpha*

Roman Numerals M

Acrophonic

X

2,000 Attic Acrophonic Numerals
阿提卡数字

Greek Alphabetic ͵Β′ 2,000
Subscript *iota* before *beta*

Roman Numerals MM

Acrophonic

XX

3,000 Attic Acrophonic Numerals
阿提卡数字

Greek Alphabetic ͵Γ′ 3,000
Subscript *iota* before *gamma*

Roman Numerals MMM

Acrophonic

XXX

4,000 Attic Acrophonic Numerals
阿提卡数字

Greek Alphabetic ͵Δ′ 4,000
Subscript *iota* before *delta*

Roman Numerals $\overline{\text{IV}}$

Acrophonic

XXXX

5,000 Attic Acrophonic Numerals
阿提卡数字

Greek Alphabetic ιE′ 5,000

Subscript *iota* before *epsilon*

Roman Numerals $\overline{\mathrm{V}}$

Acrophonic

𐅇

6,000 Attic Acrophonic Numerals
阿提卡数字

Greek Alphabetic ιF′ 6,000

Subscript *iota* before *digamma*

Roman Numerals $\overline{\mathrm{VI}}$

Acrophonic

𐅇X

7,000 Attic Acrophonic Numerals
阿提卡数字

Greek Alphabetic ιZ′ 7,000

Subscript *iota* before *zeta*

Roman Numerals $\overline{\mathrm{VII}}$

Acrophonic

𐅇XX

8,000 Attic Acrophonic Numerals
阿提卡数字

Greek Alphabetic ιH′ 8,000

Subscript *iota* before *eta*

Roman Numerals $\overline{\mathrm{VIII}}$

Acrophonic

𐅇XXX

9,000 Attic Acrophonic Numerals 阿提卡数字

Greek Alphabetic	͵Θ′ 9,000
Subscript *iota* before *theta*	
Roman Numerals	$\overline{\text{IX}}$
Acrophonic	𐅆XXXX

10,000 Attic Acrophonic Numerals 阿提卡数字

Greek Alphabetic	$\overset{\alpha}{\text{M}}$′ 10,000
M "Myriad" with small *alpha*	
Roman Numerals	$\overline{\text{X}}$
Acrophonic	M

20,000 Attic Acrophonic Numerals 阿提卡数字

Greek Alphabetic	$\overset{\beta}{\text{M}}$′ 20,000
M "Myriad" with small *beta*	
Roman Numerals	$\overline{\text{XX}}$
Acrophonic	MM

30,000 Attic Acrophonic Numerals 阿提卡数字

Greek Alphabetic	$\overset{\gamma}{\text{M}}$′ 30,000
M "Myriad" with small *gamma*	
Roman Numerals	$\overline{\text{XXX}}$
Acrophonic	MMM

40,000 Attic Acrophonic Numerals
阿提卡数字

Greek Alphabetic $\overset{\delta}{\mathrm{M}}$′ 40,000
M "Myriad" with small *delta*

Roman Numerals $\overline{\mathrm{XL}}$

Acrophonic
MMMM

50,000 Attic Acrophonic Numerals
阿提卡数字

Greek Alphabetic $\overset{\varepsilon}{\mathrm{M}}$′ 50,000
M "Myriad" with small *epsilon*

Roman Numerals $\overline{\mathrm{L}}$

Acrophonic
𐅇

60,000 Attic Acrophonic Numerals
阿提卡数字

Greek Alphabetic $\overset{\varsigma}{\mathrm{M}}$′ 60,000
M "Myriad" with small *stigma*

Roman Numerals $\overline{\mathrm{LX}}$

Acrophonic
𐅇M

70,000 Attic Acrophonic Numerals
阿提卡数字

Greek Alphabetic $\overset{\zeta}{\mathrm{M}}$′ 70,000
M "Myriad" with small *zeta*

Roman Numerals $\overline{\mathrm{LXX}}$

Acrophonic
𐅇MM

80,000 Attic Acrophonic Numerals 阿提卡数字	*90,000* Attic Acrophonic Numerals 阿提卡数字
Greek Alphabetic $\overset{\eta}{M}$′ 80,000	Greek Alphabetic $\overset{\theta}{M}$′ 90,000
M "Myriad" with small *eta*	M "Myriad" with small *theta*
Roman Numerals $\overline{LXXX}$	Roman Numerals $\overline{XC}$
Acrophonic 𐅇MMM	Acrophonic 𐅇MMMM

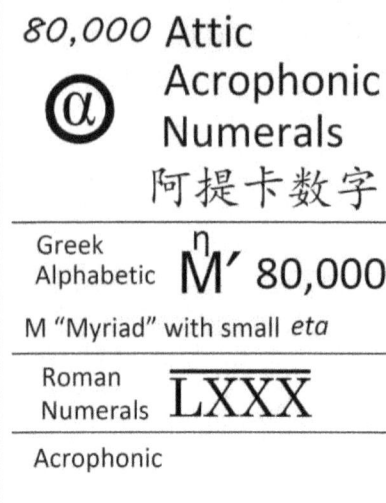

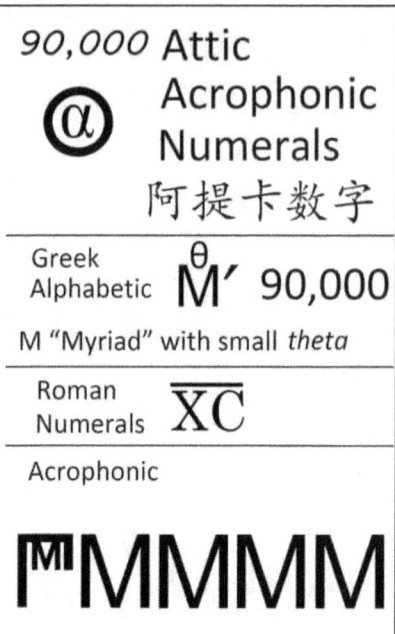

Etruscan Numerals 伊特鲁里亚

Etruscan numerals were an adaptation of Greek Attic numerals and they provided an inspiration for the design of Roman numerals. Very little evidence survives and the numbers one to six were determined by Etruscologists on the assumption that numbers on opposite sides of dice add up to seven. However, there are examples of ancient dice where this is not the case and scholars are still debating about whether *huθ* is 4 or 6 and whether *śa* is 6 or 4. Some Etruscan numbers are written as partial subtractions with 17, 18 and 19 written as three, two and one from twenty. A general agreement among scholars has assigned values to the following Etruscan numbers: θu 1; zal 2; ci 3; maχ 5; semφ 7; cezp 8; nurφ 9; śar 10; θuśar 11; zalśar12; ciśar 13; maχśar 15; ciem zaθrum 17; eslem zaθrum 18; θunem zaθrum 19; zaθrum 20; cealχ 30; muvalχ 50; semφalχ 70; cezpalχ 80; nurφalχ 90. The numbers huθ 4; śa 6; huθzar 14; śaśar 16; huθalχ 40; śealχ 60 are still under debate.

1 Etruscan Numerals
ξ 伊特鲁里亚数字

θu

I

2 Etruscan Numerals
ξ 伊特鲁里亚数字

zal

II

5 Etruscan Numerals
ξ 伊特鲁里亚数字

maχ

Λ

10 Etruscan Numerals
ξ 伊特鲁里亚数字

śar

X

11 Etruscan Numerals
伊特鲁里亚数字

θuśar

XI

12 Etruscan Numerals
伊特鲁里亚数字

zalśar

XII

15 Etruscan Numerals
伊特鲁里亚数字

Maχśar

XΛ

20 Etruscan Numerals
伊特鲁里亚数字

zaθrum

XX

50 Etruscan Numerals
伊特鲁里亚数字

muvalχ

100 Etruscan Numerals
伊特鲁里亚数字

100 Etruscan Numerals
伊特鲁里亚数字

1,000 Etruscan Numerals
伊特鲁里亚数字

Roman Numerals 古罗马数字

Roman numerals were first employed in the Old Italic script which was in use from the eighth to first century BC. They had their origin in Etruscan numerals which in turn evolved from the Greek Attic numerals. There were many variants of Roman numerals used in ancient Rome and medieval European nations. Today there is almost a standard system of Roman notation with small exceptions such as using IIII instead of IV on clock faces. This breaks the rule that there cannot be more than three identical symbols together. Another similar case could occur when writing 4,000 as MMMM instead of $\overline{\text{IV}}$. This is a common practice because most computer fonts do not have such a symbol. For 4,000 and higher values, a line above the Roman numeral indicates that it is multiplied by 1,000. $\overline{\text{V}}$ thus equals 5,000. To avoid confusion which could be caused in writing four identical symbols together, subtractive notation is used for 4, 9, 40, 90, 400 and 900. One subtracted from five is indicated by I before V which is written as IV meaning 4. Similarly 10 minus one is written as I before X or IX meaning 9. X before L means 10 less than 50 and X before C means 10 less than 100. C before D indicates 100 less than 500 and C before M indicates 100 less than 1,000.

There was no Roman numeral for zero. Medieval European scholars indicated zero by using the Latin word *nulla* meaning "none". In about 725 the Venerable Bede, an English monk, used N as a numeral to symbolize nulla "none" or nihil "nothing".

Lower case letters were developed in Europe during the Medieval Period and were popularly used in writing Roman numerals such as i, ii, iii, iv, v, vi, vii, viii, ix, x etc.

Medieval Roman Numerals A to Z and Numerology

Medieval Roman numerals, which employed most of the letters of the Roman alphabet, were used in conjunction with standard

Roman notation. Some medieval numerals replaced standard numerals; A was used for V; and Q replaced D; O was an abbreviation for XI; and F was a short form for XL. Here is a list which in conjunction with the standard Roman numerals covers every letter in the alphabet except for U and W for which I have given plausible values. This allows a person to write his or her name as a number and discover the number embedded in a name. Numerologists may be interested in this association of names with numbers. However, they are better advised to look in the section on Greek numerals in this book to find the Greek and Roman graphic equivalents of letters in the Phoenician alphabet and the numbers implied by each letter of the Roman alphabet, or Greek alphabet for that matter. Un-bracketed numbers in this table are values of Medieval Roman numerals; bracketed numbers are reasonable alternatives.

Medieval Numeral or Standard Numeral	**Number**	**Explanation** Numbers in brackets are sometimes better alternatives to Medieval Roman numerals
A	5 (1)	A resembles inverted V, standard Roman numeral for 5; Greek *alpha* **Α** equals 1.
B	300 (2, 3)	Origin unknown; Without vertical line it looks like 3; Greek *beta* **Β** = 2.
C, Ç	6, 100	Resembles Greek letter *stigma* ς = 6; standard Roman notation for 100.
D	500 (10)	Standard Roman notation for 500; resembles Attic numeral **Δ** = 10.
E	250 (5)	Origin unknown; same as Greek *epsilon* **Ε** which equals 5.
F	40 (6)	Abbreviation of English "forty"; Resembles Greek *digamma* **Ϝ** = 6.
G	400 (3)	Unknown; Greek *gamma* **Γ** equals 3.
H	200, 2 (8, 100)	H was symbol for bronze coin *dupondius* used during Roman Empire which equaled 2 *aes* or ⅛ of a *denarius*; resembles Greek *eta* **Η** = 8 and Attic numeral **Η** = 100.

I	1 (10)	Standard Roman notation for 1; resembles Greek *iota* **I** = 10.
J	1	J is a variant of Roman I and equals 1
K	151, 250 (20)	Origin unknown. Resembles Greek *kappa* **K** = 20.
L	50	Standard Roman notation for 50.
M	1,000 (10,000)	Standard Roman notation for 1,000; resembles Attic **M** = 10,000.
N	0, 90 (50)	Abbreviation of *nonaginta,* Latin for 90; Venerable Bede c. 725 AD used N for *nulla* “none” or *nihil* “nothing” meaning zero; Greek *nu* **N** = 50.
O	11 (70)	Abbreviation of *onze,* French for 11; Greek *omicron* **O** = 70.
P	400 (100)	Origin unknown; Greek letter *rho* **P** = 100.
Q	500 (90)	Redundant; D is standard Roman notation for 500; Resembles Greek *Koppa* Ϙ = 90.
R	80	Origin unknown.
S	70, 7	From Latin *septem* 7.
T	160 (300)	From Greek *tetra* 4 x 40 = 160; Greek *tau* **T** = 300.
U	(400)	Looks like Greek lower case *upsilon* **υ** = 400.
V	5	Standard Roman notation for 5.
W	(200, 800)	Looks like Greek lower case omega **ω** = 800; like Phoenician *shin* **W** which equates with Greek *sigma* **Σ** = 200 (i.e. W turned on its side with horizontal top and bottom).
X	10 (600, 1,000)	Standard Roman notation for 10; resembles Greek *chi* **X** = 600 and Attic **X** = 1,000.
Y	150 (400)	Origin unknown; Greek *upsilon* **Y** = 400.
Z	2,000 (7)	Origin unknown; Greek *zeta* **Z** = 7.

Alternatively 1 = A, J, S; 2 = B, K, T; 3 = C, L, U; 4 = D, M, V; 5 = E, N, W; 6 = F, O ,X; 7 = G, P, Y; 8 = H, Q, Z; 9 = I, R.

0 Roman Numerals
® 古罗马数字

Nihil

1 Roman Numerals
® 古罗马数字

Unus

I

2 Roman Numerals
® 古罗马数字

Duo

II

3 Roman Numerals
® 古罗马数字

Tres

III

4 Roman Numerals
Ⓡ 古罗马数字

Quattuor

IV

5 Roman Numerals
Ⓡ 古罗马数字

Quinque

V

6 Roman Numerals
Ⓡ 古罗马数字

Sex

VI

7 Roman Numerals
Ⓡ 古罗马数字

Septem

VII

8 Roman Numerals
® 古罗马数字

Octo

VIII

9 Roman Numerals
® 古罗马数字

Novem

IX

10 Roman Numerals
® 古罗马数字

Decem

X

11 Roman Numerals
® 古罗马数字

Undecem

XI

12 Roman Numerals
® 古罗马数字

Duodecim

XII

13 Roman Numerals
® 古罗马数字

Tredecim

XIII

14 Roman Numerals
® 古罗马数字

Quattuordecim

XIV

15 Roman Numerals
® 古罗马数字

Quindecim

XV

16 Roman Numerals

® 古罗马数字

Sedecim

XVI

17 Roman Numerals

® 古罗马数字

Septendecim

XVII

18 Roman Numerals

® 古罗马数字

Duodeviginti

XVIII

19 Roman Numerals

® 古罗马数字

Undeviginti

XIX

20 Roman Numerals
® 古罗马数字

Viginti

XX

30 Roman Numerals
® 古罗马数字

Triginta

XXX

40 Roman Numerals
® 古罗马数字

Quadraginta

XL

50 Roman Numerals
® 古罗马数字

Quinquaginta

L

60 Roman Numerals
® 古罗马数字

Sexaginta

LX

70 Roman Numerals
® 古罗马数字

Septuaginta

LXX

80 Roman Numerals
® 古罗马数字

Octoginta

LXXX

90 Roman Numerals
® 古罗马数字

Nonaginta

XC

100 Roman Numerals ® 古罗马数字	500 Roman Numerals ® 古罗马数字
Centum	Quingenta
C	D

1,000 Roman Numerals ® 古罗马数字	5,000 Roman Numerals ® 古罗马数字
Mille	Quinque milia
M	$\overline{V}$

10,000 Roman Numerals
® 古罗马数字

Decem milia

$\overline{X}$

25,000 Roman Numerals
® 古罗马数字

Viginti quinque-milia

$\overline{X}\overline{X}\overline{V}$

50,000 Roman Numerals
® 古罗马数字

Quinquaginta-milia

$\overline{L}$

90,000 Roman Numerals
® 古罗马数字

Nonaginta-milia

$\overline{X}\overline{C}$

100,000 Roman Numerals
® 古罗马数字

Centum milia

$\overline{\mathrm{C}}$

500,000 Roman Numerals
® 古罗马数字

Quingenta-
milia

$\overline{\mathrm{D}}$

750,000 Roman Numerals
® 古罗马数字

Septingentos-
quinquaginta-
milia

$\overline{\mathrm{DCCL}}$

1,000,000 Roman Numerals
® 古罗马数字

Decies centena-
milia

$\overline{\mathrm{M}}$

Greek Numerals 希腊数字

The letters of the Greek alphabet are also used as numerals. Most Greek letters were copied from the Phoenician alphabet of 22 consonants and many adaptations and changes were made. Some Phoenician letters were used as vowels by the Greeks and extra letters were created.

The obsolete letter Ϻ *san*, also written M, (not to be confused with M *mu*) denoted the same sound as Σ *sigma* and dropped out of use before the Greek classical period. The obsolete letters ⊏ *digamma*, Ϙ *koppa* and ϡ *sampi* were not part of the 24 letter classical Greek alphabet but were retained to represent the numbers six, 90 and 900. The shape and form of these three letters changed through the centuries. *Digamma* ⊏ or ⊏ meaning double *gamma* Γ at one time somewhat resembled an F (Ϝ, Ϝ, Ϝ). It is currently written ς and also called *stigma*. *Digamma*, which was derived from the Phoenician letter Y *waw*, had its shape radically changed to Ϝ but its position as the sixth letter in the Phoenician alphabet was at first retained in the Greek alphabet. Later, when it became obsolete, it continued to be used as a symbol for the number six. Phoenician *waw* (Y) was again borrowed by the Greeks and this time it retained its original shape but was renamed *upsilon* and placed towards the end of the alphabet. *Upsilon* symbolizes the number 400.

Greek & Latin Graphic Equivalents of Phoenician Letters

The following chart shows the Greek and Latin graphic equivalents of Phoenician letters and their Greek numerical values. Possible Latin graphic equivalents of four Greek letters are added with their Greek numerical values.

Phoe-nician	Phoenician Name	Greek Letter	Name, Value	Latin Graphical Equivalent
𐤀	Aleph	**Α** α	Alpha 1	A
𐤁	Beth	Ββ	Beta 2	B
𐤂	Gimel	**Γ** γ	Gamma 3	C, G
𐤃	Daleth	**Δ** δ	Delta 4	D
𐤄	He	**Ε** ε	Epsilon 5	E
𐤅	Waw	Ϝ Ϝ Ϝ Ϝ Ϝ ϛ ϛ	Digamma 6 Stigma 6	F, U, V, Y, W
𐤆	Zayin	Ζζ	Zeta 7	Z

𐤇	Heth	**Η** η	Eta 8	H
𐤈	Teth	**Θ** θ	Theta 9	-
𐤉	Yodh	**Ι** ι	Iota 10	I, J
𐤊	Kaph	**Κ** κ	Kappa 20	K
𐤋	Lamedh	**Λ** λ	Lambda 30	L
𐤌	Mem	**Μ** μ	Mu 40	M
𐤍	Nun	**Ν** ν	Nu 50	N
𐤎	Samekh	**Ξ** ξ	Xi 60	Possibly X
𐤏	Ayin	**Ο** ο	Omicron 70	O
𐤐	Peh	**Π** Γ π	Pi 80	P
𐤑	Sadhe	ϻ Ϻ	San	-
𐤒	Qoph	Ϙ ϟ	Koppa 90	Q

𐤓	Resh	Ρ ρ	Rho 100	R
W	Shin	Σ σ	Sigma 200	S
X	Taw	Τ τ	Tau 300	T
Y	Waw	Υ υ	Upsilon 400	U
		Φ φ	Phi 500	The right half looks like P
		Χ χ	Chi 600	X
		Ψ ψ	Psi 700	-
		Ω ω	Omega 800	O, Q, W
		ϡ	Sampi 900	-

The first nine letters of the old Ionic alphabet were used as symbols for the numbers one to nine. Then the next nine letters of the Ionic alphabet were used as symbols for multiple of 10 from 10 to 90. Each of the hundreds from 100 to 900 was also assigned a letter of the alphabet. However, the Greek alphabet has 24 letters and it was therefore necessary to use the obsolete letters *digamma*, *koppa* and *sampi* to make 27. This system of numbering was probably in use by the 5th century BC. It was long resisted by Athens where full adoption was eventually achieved by 50 AD. The letters used in a

Greek number are arranged from highest to lowest and their numeric values are added together. To distinguish a Greek number from other words and letters, a small mark known as a *keraia* is placed at its upper right.

For numbers from 1,000 to 9,000 a superscript or subscript *iota* ι is placed before the letters symbolizing one to nine. For example a small superscript or subscript *iota* ι before *beta* Β symbolizes 2,000. See further examples in this section. Myriad notation was used in ancient Greek for numbers greater than 9,999. Small or lower case Greek letters representing a number up to 9,999 were written above the symbol Μ for myriad and multiplied by 10,000. Thus writing a small or lower case *gamma* γ above Μ represented a value of 30,000. See further examples in this section. It could be difficult to write a long number of many small letters above Μ and in this case the letters were written in upper case in front of the Μ. See examples in this section.

0 G Greek Numerals 希腊数字

0 mēdén μηδέν

1 G Greek Numerals 希腊数字

1 ena ένα

Α' alpha

α' alpha

2 Ⓖ Greek Numerals 希腊数字

2 dyo δυο

Β' beta

β' beta

3 Ⓖ Greek Numerals 希腊数字

3 tria τρία

Γ' gamma

γ' gamma

4 Ⓖ Greek Numerals 希腊数字

4 tessera τέσσερα

Δ' delta

δ' delta

5 Ⓖ Greek Numerals 希腊数字

5 pente πέντε

Ε' epsilon

ε' epsilon

6 Ⓖ Greek Numerals 希腊数字

6 eksi
έξι

Ϝ' digamma

ς' stigma

7 Ⓖ Greek Numerals 希腊数字

7 epta
επτά

Ζ' zeta

ζ' zeta

8 Ⓖ Greek Numerals 希腊数字

8 oktw
οκτώ

Η' eta

η' eta

9 Ⓖ Greek Numerals 希腊数字

9 ennea
εννέα

Θ' theta

θ' theta

10 Ⓖ Greek Numerals 希腊数字

10 deka δέκα

Ι'   iota

ι'   iota

11 Ⓖ Greek Numerals 希腊数字

11 enteka έντεκα

ΙΑ'

ια'

12 Ⓖ Greek Numerals 希腊数字

12 dwdeka δώδεκα

ΙΒ'

ιβ'

13 Ⓖ Greek Numerals 希腊数字

13 dekatria δεκατρία

ΙΓ'

ιγ'

14 Ⓖ	Greek Numerals 希腊数字
14	dekatessera δεκατέσσερα
ΙΔ' ιδ'	

15 Ⓖ	Greek Numerals 希腊数字
15	dekapente δεκαπέντε
ΙΕ' ιε'	

16 Ⓖ	Greek Numerals 希腊数字
16	dekaeksi δεκαέξι
ΙϜ' ις'	

17 Ⓖ	Greek Numerals 希腊数字
17	dekaepta δεκαεπτά
ΙΖ' ιζ'	

18 Greek Numerals 希腊数字

18 dekaoxtw δεκαοχτώ

ΙΗ'

ιη'

19 Greek Numerals 希腊数字

19 dekaennea δεκαεννέα

ΙΘ'

ιθ'

20 Greek Numerals 希腊数字

20 eikosi είκοσι

Κ' kappa

κ' kappa

30 Greek Numerals 希腊数字

30 τριαντα

Λ' lambda

λ' lambda

40 Ⓖ Greek Numerals 希腊数字
40 σαραντα
M' mu
μ' mu

50 Ⓖ Greek Numerals 希腊数字
50 πενηντα
N' nu
ν' nu

60 Ⓖ Greek Numerals 希腊数字
60 εξηντα
Ξ' xi
ξ' xi

70 Ⓖ Greek Numerals 希腊数字
70 εβδομηντα
O' omicron
ο' omicron

80 Greek Numerals 希腊数字

80 ογδοντα

Π' pi

π' pi

90 Greek Numerals 希腊数字

90 ενενηντα

Ϙ' koppa

ϟ' koppa

100 Greek Numerals 希腊数字

100 εκατο

Ρ' rho

ρ' rho

200 Greek Numerals 希腊数字

200 διακοσια

Σ' sigma

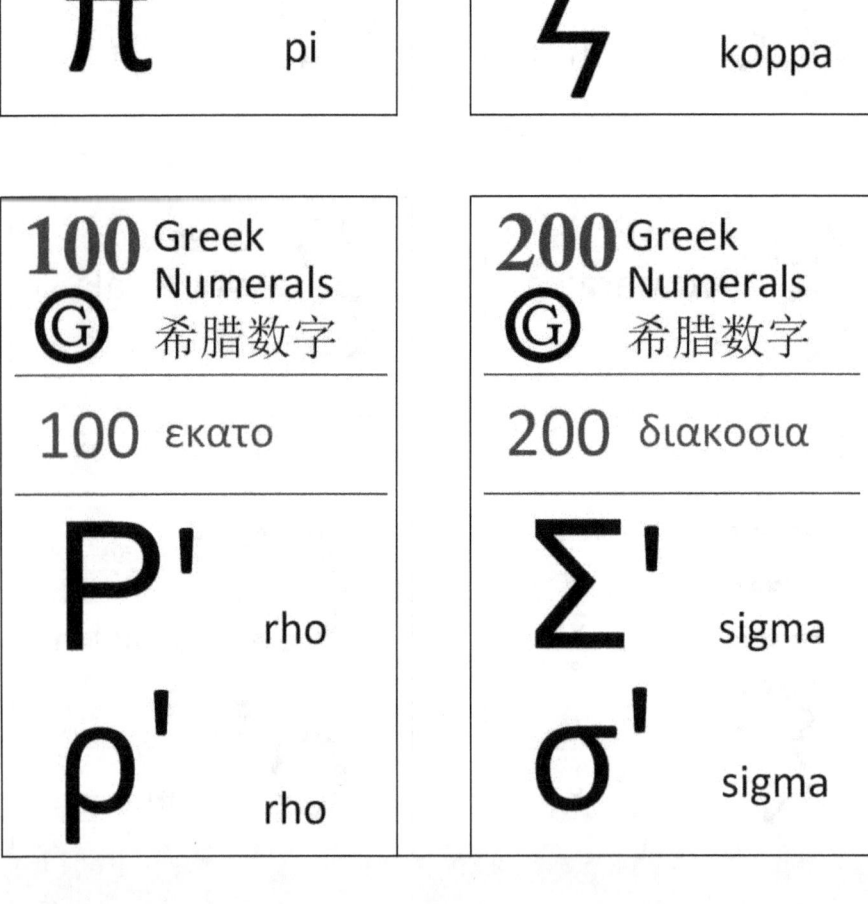

σ' sigma

300 Greek Numerals 希腊数字
300 τριακοσια
Τ' tau
τ' tau

400 Greek Numerals 希腊数字
400 τετρακοσια
Υ' upsilon
υ' upsilon

500 Greek Numerals 希腊数字
500 πεντακοσια
Φ' phi
φ' phi

600 Greek Numerals 希腊数字
600 εξακοσια
Χ' chi
χ' chi

700 Greek Numerals 希腊数字

700 επτακοσια

Ψʹ psi

ψʹ psi

800 Greek Numerals 希腊数字

800 οκτακοσια

Ωʹ omega

ωʹ omega

900 Greek Numerals 希腊数字

900 εννιακοσια

ϡʹ sampi

1,000 Greek Numerals 希腊数字

1,000 xilia χίλια

͵Αʹ

͵αʹ

2,000 Greek Numerals 希腊数字

2,000
δύο χιλιάδες

͵Β'

͵β'

4,000 Greek Numerals 希腊数字

4,000
τέσσερις χιλιάδες

͵Δ

͵δ'

5,000 Greek Numerals 希腊数字

5,000
πέντε χιλιάδες

͵Ε'

͵ε'

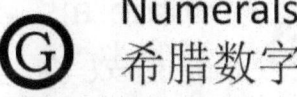

10,000 Greek Numerals 希腊数字

10,000
δέκα χιλιάδες

$\overset{\alpha}{M}$'

Explanation

α M $\overset{\alpha}{M}$'
1 x 10,000 = 10,000

40,000 Greek Numerals 希腊数字

40,000
σαράντα χιλιάδες

δ
M'

Explanation

δ M M'(δ)
4 x 10,000 = 40,000

100,000 Greek Numerals 希腊数字

100,000
εκατό χιλιάδες

ι
M'

Explanation

ι M M'(ι)
10 x 10,000 = 100,000

1,000,000 Greek Numerals 希腊数字

ekatommyrio
εκατομμύριο

ρ
M'

Explanation

ρ M
100 x 10,000 =
= 1,000,000

1,570,000 Greek Numerals 希腊数字

1,570,000

ρνζ
M'

Explanation

ρ ν ζ M
100 + 50 + 7 x 10,000
= 1,570,000

2,000,000 Greek Numerals 希腊数字	98,326,793 Greek Numerals 希腊数字
2,000,000 δύο εκατομμύρια	98,326,793
σ M' Explanation σ M 200 x 10,000 = = 2,000,000	͵ΘΩΛΒΜ ͵FΨϘΓ Explanation ͵Θ Ω Λ Β ͵ΘΩΛΒ 9,000 + 800 + 30 + 2 = 9,832 ͵ΘΩΛΒ Μ ͵ΘΩΛΒΜ 9,832 x 10,000 = 98,320,000 ͵F Ψ Ϙ Γ ͵FΨϘΓ 6,000 + 700 + 90 + 3 = 6,793 98,320,000 + 6,793 = 98,326,793

Amharic Numerals 阿姆哈拉数字

Amharic is the national language of Ethiopia. There are almost 30 million speakers making it the second largest Semitic language. Only Arabic is larger. It is spoken in North and Central Ethiopia, Eritrea, as well as in parts of Egypt and Israel. The script and numerals are written left to right. There are 33 basic consonants each one combining with one of the seven vowels forming a script of syllables. Amharic is based on Ethiopia's classical language Ge'ez which has inscriptions dating back to the 5th century BC and at that time it used only consonants. Vowels were introduced in the 3rd century AD. Ge'ez is still used as a liturgical language by Orthodox Christians, Afro-Jamaican Rastafarians and Ethiopia's Jewish community. Much of the script and some numerals were borrowed from Greek through Coptic. Each numeral has a line above and a line below. Amharic lacks a traditional numeral for zero and numbers start with one ፩ (pronounced *and* አንድ). Nowadays a circular zero from

Western Indo-Arabic notation is used without a line above or below to avoid confusion with number four ፬. There are separate numerals for 10, 20, 30, up to 90. There is no numeral for 1,000 and instead the alphabet syllable *shih* ሺ is often used. *Shih* is written without a line above and below indicating that it is not a traditional numeral. Alternatively 1,000 can be written as 10 times 100 ፲፻. The number 2,200 can be written ፳፪፻ (22 x 100) to avoid using a non-numeric alphabet syllable. The numeral for 100 is ፻ and it is doubled for 10,000 ፼ (i.e. 100 x 100). For mathematics, science and modern writing, the Western Indo-Arabic numerals are used with the traditional script.

2 Amharic Numerals 阿姆哈拉数字

hulät

ሁለት

፪

3 Amharic Numerals 阿姆哈拉数字

sost

ሦስት

፫

4 Amharic Numerals 阿姆哈拉数字

arat

አራት

፬

5 Amharic Numerals 阿姆哈拉数字

ammәst

አምስት

፭

6 Amharic Numerals
阿姆哈拉数字

səddəst
ስድስት

፮

7 Amharic Numerals
阿姆哈拉数字

säbat
ሰባት

፯

8 Amharic Numerals
阿姆哈拉数字

səmmənt
ስምንት

፰

9 Amharic Numerals
阿姆哈拉数字

zäṭäñ
ዘጠኝ

፱

10 Amharic Numerals 阿姆哈拉数字

asər
አስር （አሥር）

፲

11 Amharic Numerals 阿姆哈拉数字

asər and
አሥር አንድ

፲፩

12 Amharic Numerals 阿姆哈拉数字

asər hulät
አሥር ሁለት

፲፪

13 Amharic Numerals 阿姆哈拉数字

asər sost
አሥር ሦስት

፲፫

14 Amharic Numerals 阿姆哈拉数字

asər arat
አሥር አራት

፲፬

15 Amharic Numerals 阿姆哈拉数字

asər amməst
አሥር አምስት

፲፭

16 Amharic Numerals 阿姆哈拉数字

asər səddəst
አሥር ስድስት

፲፮

17 Amharic Numerals 阿姆哈拉数字

asər säbat
አሥር ሰባት

፲፯

18 Amharic Numerals 阿姆哈拉数字	19 Amharic Numerals 阿姆哈拉数字
asər səmmənt አሥር ስምንት	asər zäṭäñ አሥር ዘጠኝ
፲፰	፲፱

20 Amharic Numerals 阿姆哈拉数字	30 Amharic Numerals 阿姆哈拉数字
haya ሀያ （ሃያ）	sälasa ሰላሳ
፳	፴

40 Amharic Numerals 阿姆哈拉数字

arba
አርባ

፵

50 Amharic Numerals 阿姆哈拉数字

hamsa
ሀምሳ

፶

60 Amharic Numerals 阿姆哈拉数字

səlsa
ስልሳ

፷

70 Amharic Numerals 阿姆哈拉数字

säba
ሰባ

፸

80 Amharic Numerals
阿姆哈拉数字

sämaňa
ሰማንያ

፹

90 Amharic Numerals
阿姆哈拉数字

zätena
ዘጠና

፺

100 Amharic Numerals
阿姆哈拉数字

mäto (meto)
መቶ

፻

200 Amharic Numerals
阿姆哈拉数字

hulät mäto
ሁለት መቶ

፪፻

300 Amharic Numerals
阿姆哈拉数字

sost mäto
ሦስት መቶ

፫፻

400 Amharic Numerals
阿姆哈拉数字

arat mäto
አራት መቶ

፬፻

500 Amharic Numerals
阿姆哈拉数字

amməst mäto
አምስት መቶ

፭፻

600 Amharic Numerals
阿姆哈拉数字

səddəst mäto
ስድስት መቶ

፮፻

700 Amharic Numerals
阿姆哈拉数字

säbat mäto
ሰባት መቶ

፯፻

800 Amharic Numerals
阿姆哈拉数字

səmmənt mäto
ስምንት መቶ

፰፻

900 Amharic Numerals
阿姆哈拉数字

zäṭäñ mäto
ዘጠኝ መቶ

፱፻

1,000 Amharic Numerals
阿姆哈拉数字

ši (shih)
ሺህ

ሺ

1,000 Amharic Numerals
阿姆哈拉数字

asər mäto (meto)
አሥር መቶ

፲፻

2,200 Amharic Numerals
阿姆哈拉数字

haya hulät mäto
ሀያ ሁለት መቶ

፳፪፻

10,000 Amharic Numerals
阿姆哈拉数字

asir shih (asir ši)
አስር ሺህ

፻፻

100,000 Amharic Numerals
阿姆哈拉数字

mäto ši (meto shih)
መቶ ሺህ

፲፻፻

1,000,000 Amharic Numerals
阿姆哈拉数字

100 x 100 x 100 =
1,000,000

፻፼

10,000,000 Amharic Numerals
阿姆哈拉数字

10 x 100 x 100 x 100 =
10,000,000

፲፻፼

100,000,000 Amharic Numerals
阿姆哈拉数字

100 x 100 x 100 x 100 =
100,000,000

፼፼

Hebrew Numerals 希伯来

Numbering systems used by the Hebrews as early as about 800 BC were derived from the Aramaic and Phoenician systems which had their origin in Egyptian Hieratic numerals. The current Hebrew numbering system uses the letters of the Hebrew alphabet as numerals. It was adapted from the Greek Ionian numerals in the late 2nd century BC. There was no symbol for zero and the units 1 to 9 were each assigned a separate letter of the alphabet. The tens, 10, 20, 30 up to 90, were also given a separate letter followed by the hundreds from 100 up to 400. Then numbers from 100 to 400 were added together to obtain values for 500 to 900. Alternatively, an extension of the numeral system covers numbers 500 to 900 and this is displayed here together with the added numbers. Numbers must agree in gender with the noun they describe. Feminine gender is used in cases where no gender is obvious such as in telephone numbers. There is no gender for numbers 20 and above.

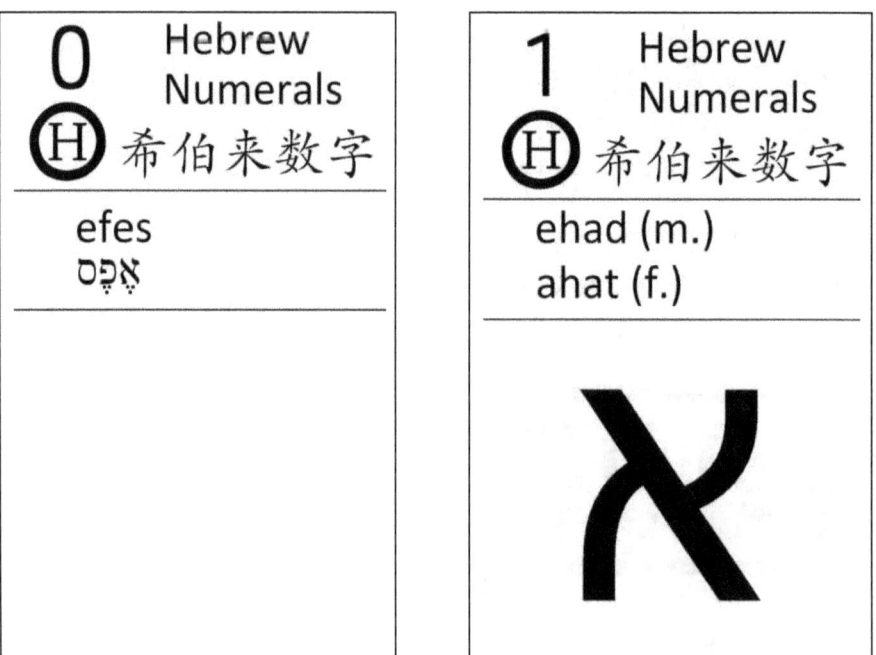

0 Hebrew Numerals (H) 希伯来数字	1 Hebrew Numerals (H) 希伯来数字
efes אֶפֶס	ehad (m.) ahat (f.)
	א

2 Hebrew Numerals

Ⓗ 希伯来数字

shnayim (m.)
shtayim (f.)

ב

3 Hebrew Numerals

Ⓗ 希伯来数字

shlosha (m.)
shalosh (f.)

ג

4 Hebrew Numerals

Ⓗ 希伯来数字

arba'a (m.)
arba' (f.)

ד

5 Hebrew Numerals

Ⓗ 希伯来数字

hamisha (m.)
hamesh (f.)

ה

6 Hebrew Numerals
Ⓗ 希伯来数字

shisha (m.)
shesh (f.)

ו

7 Hebrew Numerals
Ⓗ 希伯来数字

shiv'a (m.)
sheva' (f.)

ז

8 Hebrew Numerals
Ⓗ 希伯来数字

shmona (m.)
shmone (f.)

ח

9 Hebrew Numerals
Ⓗ 希伯来数字

tish'a (m.)
tesha' (f.)

ט

10 Hebrew Numerals (H) 希伯来数字	11 Hebrew Numerals (H) 希伯来数字
'assara (m.) 'eser (f.)	achad-'asar (m.) achat-'esre (f.)
י	יא

12 Hebrew Numerals (H) 希伯来数字	13 Hebrew Numerals (H) 希伯来数字
shneyim-'asar (m.) shteyim-'esre (f.)	shlosha-'asar (m.) shlosh-'esre (f.)
יב	יג

14 Hebrew Numerals
Ⓗ 希伯来数字

arba'a-'asar (m.)
arba'-'esre (f.)

יד

15 Hebrew Numerals
Ⓗ 希伯来数字

hamisha-'asar (m.)
hamesh-'esre (f.)

ט״ו

16 Hebrew Numerals
Ⓗ 希伯来数字

shisha-'asar (m.)
shesh-'esre (f.)

ט״ז

17 Hebrew Numerals
Ⓗ 希伯来数字

shiv'a-'asar (m.)
shva'-'esre (f.)

יז

18 Hebrew Numerals

H 希伯来数字

shmona-'asar (m.)
shmone-'esre (f.)

יח

19 Hebrew Numerals

H 希伯来数字

tish'a-'asar (m.)
tesha'-'esre (f.)

יט

20 Hebrew Numerals

H 希伯来数字

'esrim
עֶשְׂרִים

כ

30 Hebrew Numerals

H 希伯来数字

shloshim
שְׁלוֹשִׁים

ל

40 Hebrew Numerals

Ⓗ 希伯来数字

arba'im

אַרְבָּעִים

מ

50 Hebrew Numerals

Ⓗ 希伯来数字

hamishim

חֲמִשִּׁים

נ

60 Hebrew Numerals

Ⓗ 希伯来数字

shishim

שִׁשִּׁים

ס

70 Hebrew Numerals

Ⓗ 希伯来数字

shiv'im

שִׁבְעִים

ע

80 Hebrew Numerals
H 希伯来数字

shmonim
שְׁמוֹנִים

פ

90 Hebrew Numerals
H 希伯来数字

tish'im
תִּשְׁעִים

צ

100 Hebrew Numerals
H 希伯来数字

mea
מֵאָה

ק

200 Hebrew Numerals
H 希伯来数字

matayim
מָאתַיִם

ר

300 Hebrew Numerals
Ⓗ 希伯来数字

shlosh meot
מֵאוֹת שְׁלוֹשׁ

ש

400 Hebrew Numerals
Ⓗ 希伯来数字

arba' meot
מֵאוֹת אַרְבַּע

ת

500 Hebrew Numerals
Ⓗ 希伯来数字

hamesh meot
מֵאוֹת חֲמֵשׁ

ך

ת"ק

600 Hebrew Numerals
Ⓗ 希伯来数字

shesh meot
מֵאוֹת שֵׁשׁ

ם

ת"ר

700 Hebrew Numerals
Ⓗ 希伯来数字

shva meot
מֵאות שְׁבַע

ן

ת״ש

800 Hebrew Numerals
Ⓗ 希伯来数字

shmone meot
מֵאות שְׁמוֹנֶה

ף

ת״ת

900 Hebrew Numerals
Ⓗ 希伯来数字

tsha' meot
מֵאות תְּשַׁע

ץ

תת״ק

1,000 Hebrew Numerals
Ⓗ 希伯来数字

elef
אֶלֶף

א׳

2,000 Hebrew Numerals
Ⓗ 希伯来数字

alpaim
אַלְפַּיִם

ב׳

5,000 Hebrew Numerals
Ⓗ 希伯来数字

hameshet alafim
אֲלָפִים חֲמֵשֶׁת

ה׳

10,000 Hebrew Numerals
Ⓗ 希伯来数字

aseret alafim
אֲלָפִים עֲשֶׂרֶת

א׳י

100,000 Hebrew Numerals
Ⓗ 希伯来数字

mea elef
אֶלֶף מֵאָה

א׳ק

1,000,000 Hebrew Numerals

Ⓗ 希伯来数字

miliyon

מליון

א'א'

The Greek, Hebrew and Arabic Numerals and their Origin in the Phoenician Alphabet

The Hebrew numbering system uses the letters of the Hebrew alphabet as numerals. It was adapted from the Greek Ionian numerals in the late 2nd century BC. The Greek alphabet on which the numerals are based was adapted from the Phoenician alphabet and these origins are shown in the table below. Greek numbers differ from Hebrew numbers after 80 because the Greek letter *san* derived from the Phoenician letter *sadhe* was abandoned before the classical period. The Phoenician alphabet has only 22 letters, not enough for the Greeks and Arabs who each added six extra letters to their respective alphabets. The Arabs used their *abjad* (alphabetic) numerals for mathematics and accounting until the 8th century when they adopted Eastern Indo-Arabic numerals.

Phoenician Letter	Phoenician Name, Number	Greek Letter, Value and Name	Hebrew Letter	Hebrew Value & Name	Arabic Letter	Arabic Name and Value
𐤀	Aleph 1	Α α 1 Alpha	א	1 Alef	ا	Alif 1
𐤁	Beth 2	Β β 2 Beta	ב	2 Bet	ب	Beh 2
𐤂	Gimel 3	Γ γ 3 Gamma	ג	3 Gimel	ج	Jim 3
𐤃	Daleth 4	Δ δ 4 Delta	ד	4 Dalet	د	Dal 4
𐤄	He 5	Ε ε 5 Epsilon	ה	5 He	ه	Heh 5
𐤅	Waw 6	Ϝ ϝ Digamma Ϛ ϛ 6 Stigma	ו	6 Vav	و	Waw 6
𐤆	Zayin 7	Ζ ζ 7 Zeta	ז	7 Zayin	ز	Zin 7
𐤇	Heth 8	Η η 8 Eta	ח	8 Het	ح	Ha 8
𐤈	Teth 9	Θ θ 9 Theta	ט	9 Tet	ط	Ta 9

𐤉	Yodh 10	Ι ι 10 Iota	י	10 Yod	ي	Yeh 10
𐤊	Kaph 11	Κ κ Kappa 20	כ, ך	20 Kaf	ك	Kaf 20
𐤋	Lamedh 12	Λ λ Lambda 30	ל	30 Lamed	ل	Lam 30
𐤌	Mem 13	Μ μ Mu 40	מ, ם	40 Mem	م	Mim 40
𐤍	Nun 14	Ν ν Nu 50	נ, ן	50 Nun	ن	Nun 50
𐤎	Samekh 15	Ξ ξ Xi 60	ס	60 Samekh	س	Sin 60
𐤏	Ayin 16	Ο ο Omicron 70	ע	70 Ayin	ع	‘Ain 70
𐤐	Peh 17	ΠΓπ Pi 80	פ, ף	80 Pe	ف	Feh 80

𐤑	Sadhe 18	Ϻ ϻ San -	צ, ץ	90 Tsadi	ص	Sad 90
𐤒	Qoph 19	Ϙ ϟ Koppa 90	ק	100 Qof	ق	Qaf 100
𐤓	Resh 20	Ρ ρ Rho 100	ר	200 Resh	ر	Ra 200
W	Shin 21	Σ σ Sigma 200	ש	300 Shin	ش	Shin 300
X	Taw 22	Τ τ Tau 300	ת	400 Tav	ت	Tch 400
Y	Waw (repeat of 6)	Υ υ Upsilon 400	ך	500 Final Kaf	ث	Theh 500
		Φ φ Phi 500	ם	600 Final Mem	خ	Kha 600
		Χ χ Chi 600	ן	700 Final Nun	ذ	Dhal 700
		Ψ ψ Psi 700	ף	800 Final Pe	ض	Dad 800

		Ω ω Omega 800	ץ	900 Final Tsadi	ظ	Dha 900
		ϡ Sampi 900			غ	Ghain 1,000

The Hebrew letters and numerals are equivalent to the Arabic *abjad* letters and numerals up to 400. The Hebrew alphabet, like its Phoenician parent, ended with the 22nd letter *Tav* but the final forms of *Kaf* ך, *Mem* ם, *Nun* ן, *Pe* ף and *Tsadi* ץ were added to bring the number to 27 and allow for representation of numerals up to 900. However, 400 to 900 are usually represented by the juxtapositions ת"ק (100 + 400), ת"ר (200 + 400), ת"ש (300 + 400), ת"ת (400 + 400) and תת"ק (100 + 400 + 400).

A geresh placed after a Hebrew number multiplies its value by 1,000. For example, the letter *He* ה represents the number 5 and placing a geresh ׳ after it, makes the number 5,000 ה׳. This is similar to placing a small subscript or superscript *iota* before *epsilon* in Greek ͵Ε' or placing a line over the Latin V to make 5,000 $\overline{V}$.

The use of an alphanumeric code to assign numeric values to words and phrases is known in Hebrew as *gematria* and is part of the mystical *kabbalistic* practice. This has led to the number 18 being considered a lucky number in Jewish culture. Gematria is also used in English numerology and this practice in Greek is known as *isopsephy*. In Arabic the *abjad* numerals are used for this purpose.

Arabic Abjad Numerals 字母顺序的阿拉伯数字

The system of Arabic *Abjad* numerals assigns numerical values to the 28 letters of the Arabic alphabet. Mathematics and accounting depended on this system before the introduction of the Indo-Arabic numerals in the eighth century. The first 9 letters of the Arabic alphabet *alif*, *beh*, *jim* etc. were assigned to the units 1 to 9 and the next nine letters symbolized the tens from 10 to 90. Then another nine letters were assigned to the hundreds with a final letter representing 1,000. The first 22 letters of the Arabic alphabet have their origin in the Phoenician alphabet of 22 letters which had developed before the 10th century BC. An additional 6 letters were added to make 28 and represent numbers up to 1,000. Nowadays the *abjad* numerals are used for numbering lists, much as A, B, C, D or i, ii, iii, iv etc. are used in English. They are also used in numerology to assign numbers to Arabic names and words.

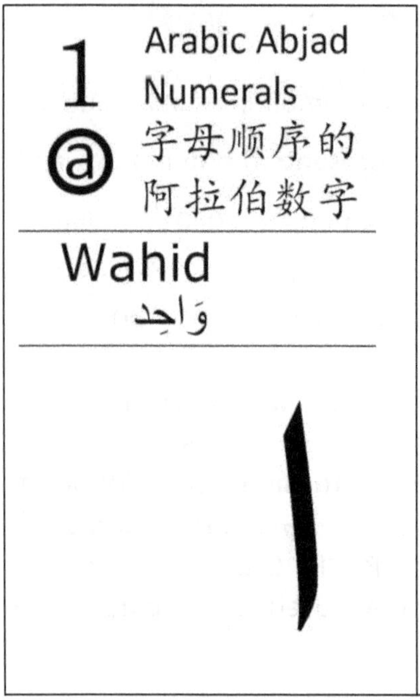

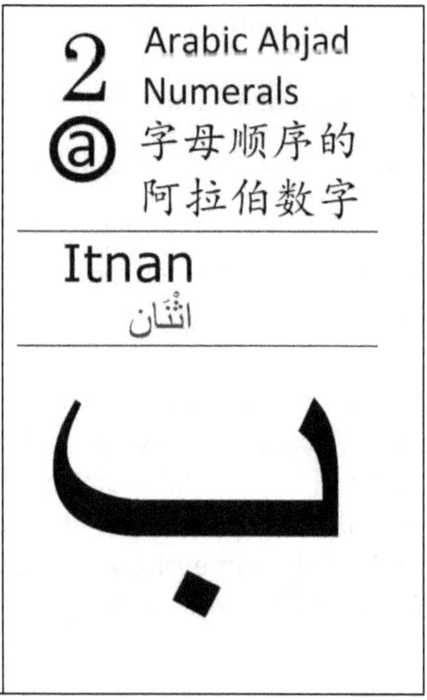

3 Arabic Abjad Numerals

@ 字母顺序的阿拉伯数字

Talatah

ثَلَاثَة

ج

4 Arabic Abjad Numerals

@ 字母顺序的阿拉伯数字

Arab'ah

أَرْبَعَة

د

5 Arabic Abjad Numerals

@ 字母顺序的阿拉伯数字

Hamsah

خَمْسَة

ه

6 Arabic Abjad Numerals

@ 字母顺序的阿拉伯数字

Sittah

سِتَّة

و

7 Arabic Abjad Numerals
@ 字母顺序的阿拉伯数字

Sab'ah
سَبْعَة

ز

8 Arabic Abjad Numerals
@ 字母顺序的阿拉伯数字

Tamaniyyah
ثَمَانِيَة

ح

9 Arabic Abjad Numerals
@ 字母顺序的阿拉伯数字

Tis'ah
تِسْعَة

ط

10 Arabic Abjad Numerals
@ 字母顺序的阿拉伯数字

Asarah
عَشْرَة

ي

20 Arabic Abjad Numerals
@ 字母顺序的阿拉伯数字

Ishrin
عِشْرُوْن

ك

30 Arabic Abjad Numerals
@ 字母顺序的阿拉伯数字

Talatin
ثَلاثونَ

ل

40 Arabic Abjad Numerals
@ 字母顺序的阿拉伯数字

Arba’ain
أَرْبَعونَ

م

50 Arabic Abjad Numerals
@ 字母顺序的阿拉伯数字

Kamisin
خَمْسونَ

ن

60 Arabic Abjad Numerals
@ 字母顺序的阿拉伯数字

Sitin
سِتّونَ

س

70 Arabic Abjad Numerals
@ 字母顺序的阿拉伯数字

Saba'ain
سَبْعونَ

ع

80 Arabic Abjad Numerals
@ 字母顺序的阿拉伯数字

Tamanin
ثَمانونَ

ف

90 Arabic Abjad Numerals
@ 字母顺序的阿拉伯数字

Tisain
تِسْعونَ

ص

100 ⓐ Arabic Abjad Numerals
字母顺序的阿拉伯数字

Mia'a
مِائَة

ق

---

200 ⓐ Arabic Abjad Numerals
字母顺序的阿拉伯数字

Mi'atān
مِائَتَانِ

ر

---

300 ⓐ Arabic Abjad Numerals
字母顺序的阿拉伯数字

Thalāth mi'ah
ثَلاثمِـائة

ش

400 ⓐ Arabic Abjad Numerals
字母顺序的阿拉伯数字

Arba' mi'ah
أَرْبَعمِائة

ت

500 Arabic Abjad Numerals
ⓐ 字母顺序的阿拉伯数字

Khamsu mi'ah
خَمْسمِائة

ث

600 Arabic Abjad Numerals
ⓐ 字母顺序的阿拉伯数字

Sitta mi'ah
سِتّمِائة

خ

700 Arabic Abjad Numerals
ⓐ 字母顺序的阿拉伯数字

Sab'a mi'ah
سَبْعمِائة

ذ

800 Arabic Abjad Numerals
ⓐ 字母顺序的阿拉伯数字

Thamānii mi'ah
ثَمانيمِائة

ض

900 Arabic Abjad Numerals
ⓐ 字母顺序的阿拉伯数字

Tis'a mi'ah
تِسْعمِائة

ظ

1,000 Arabic Abjad Numerals
ⓐ 字母顺序的阿拉伯数字

Aalf
أَلْف

غ

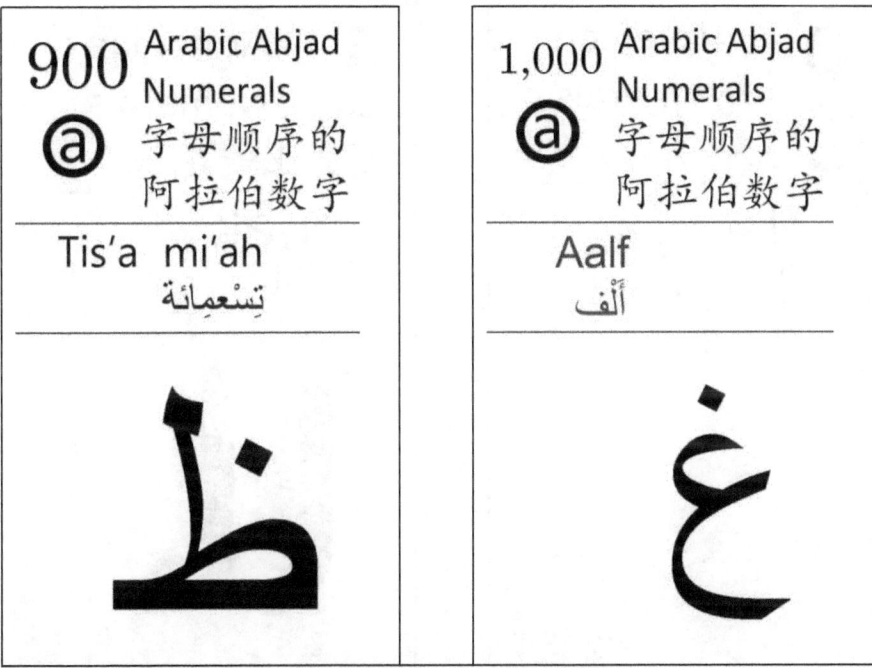

Armenian Numerals 亚美尼亚数字

The Armenian alphabet was introduced in 405 AD by Mesrop Mashtots. The upper case letters are used to write numerals except for the two final letters which were later additions and do not have numerical values. Armenian is written left to right and so are the numerals. There was no symbol for zero and the system of numerals was similar to that employed for Greek and Hebrew. In modern Armenia the Western Indo-Arabic numerals are mostly used but traditional alphabetic numerals are employed for lists, much as in the English usage of A, B, C, or i, ii, iii, etc. Numerals are traditionally written larger to smaller in decreasing value order and added together to obtain the intended number. A line drawn over a number multiplies it by 10,000 and this system is used for writing numbers greater than 9,999.

0 Armenian Numerals
Ⓕ 亚美尼亚数字

zro
զրո

---

1 Armenian Numerals
Ⓕ 亚美尼亚数字

mek
մեկ

Ա

---

2 Armenian Numerals
Ⓕ 亚美尼亚数字

yerkou
երկու

Բ

---

3 Armenian Numerals
Ⓕ 亚美尼亚数字

yereḳ
երեք

Գ

4 Armenian Numerals
Ⓕ 亚美尼亚数字

č̣ors
չորս

Դ

5 Armenian Numerals
Ⓕ 亚美尼亚数字

hing
հինգ

Ե

6 Armenian Numerals
Ⓕ 亚美尼亚数字

veç̣
վեց

Զ

7 Armenian Numerals
Ⓕ 亚美尼亚数字

yoṭ
յոթ

Է

8 Armenian Numerals
Ⓕ 亚美尼亚数字

outʻ
ութ

Ը

9 Armenian Numerals
Ⓕ 亚美尼亚数字

inə
ինը

Թ

10 Armenian Numerals
Ⓕ 亚美尼亚数字

tasə
տասը

Ժ

11 Armenian Numerals
Ⓕ 亚美尼亚数字

tasnmek
տասնմեկ

ԺԱ

12 Armenian Numerals

Ⓕ 亚美尼亚数字

tasnerku

տասներկու

ԺԲ

13 Armenian Numerals

Ⓕ 亚美尼亚数字

tasnereḳ

տասներեք

ԺԳ

14 Armenian Numerals

Ⓕ 亚美尼亚数字

tasnçors

տասնչորս

ԺԴ

15 Armenian Numerals

Ⓕ 亚美尼亚数字

tasnhing

տասնհինգ

ԺԵ

16 Armenian Numerals

Ⓕ 亚美尼亚数字

tasnveç

տասնվեց

ԺԶ

17 Armenian Numerals

Ⓕ 亚美尼亚数字

tasnyoṭ

տասնյոթ

ԺԷ

18 Armenian Numerals

Ⓕ 亚美尼亚数字

tasnouṭ

տասնութ

ԺԸ

19 Armenian Numerals

Ⓕ 亚美尼亚数字

tasninə

տասնինը

ԺԹ

20 Armenian Numerals
Ⓕ 亚美尼亚数字

k’san
քսան

Ի

30 Armenian Numerals
Ⓕ 亚美尼亚数字

eresown
երեսուն

Լ

40 Armenian Numerals
Ⓕ 亚美尼亚数字

k’arasown
քառասուն

Խ

50 Armenian Numerals
Ⓕ 亚美尼亚数字

hisown
հիսուն

Ծ

60 Armenian Numerals
Ⓕ 亚美尼亚数字

vat'sown
վաթսուն

Կ

70 Armenian Numerals
Ⓕ 亚美尼亚数字

yot'anasown
յոթանասուն

Հ

80 Armenian Numerals
Ⓕ 亚美尼亚数字

owt'sown
ութսուն

Ձ

90 Armenian Numerals
Ⓕ 亚美尼亚数字

innsown
իննսուն

Ղ

100 Armenian Numerals
Ⓕ 亚美尼亚数字

haryowr
հարյուր

Ճ

200 Armenian Numerals
Ⓕ 亚美尼亚数字

yerkoo hyeroor
երկու հարյուր

Մ

300 Armenian Numerals
Ⓕ 亚美尼亚数字

yehreq hahryoor
երեք հարյուր

Յ

400 Armenian Numerals
Ⓕ 亚美尼亚数字

chorse hahryoor
չորս հարյուր

Ն

500 Armenian Numerals
Ⓕ 亚美尼亚数字

heeng hahryoor
հինգ հարյուր

Ճ

600 Armenian Numerals
Ⓕ 亚美尼亚数字

vehts hahryoor
վեց հարյուր

Մ

700 Armenian Numerals
Ⓕ 亚美尼亚数字

yote hahryoor
յոթ հարյուր

Յ

800 Armenian Numerals
Ⓕ 亚美尼亚数字

oote hahryoor
ութ հարյուր

Ն

900 Armenian Numerals
Ⓕ 亚美尼亚数字

eene hahryoor
ինը հարյուր

Ջ

1,000 Armenian Numerals
Ⓕ 亚美尼亚数字

hazar
հազար

Ռ

2,000 Armenian Numerals
Ⓕ 亚美尼亚数字

yerkou hahzahr
երկու հազար

Ս

3,000 Armenian Numerals
Ⓕ 亚美尼亚数字

yehreq hahzahr
երեք հազար

Վ

4,000 Armenian Numerals
Ⓕ 亚美尼亚数字

chorse hahzahr
չորս հազար

Տ

5,000 Armenian Numerals
Ⓕ 亚美尼亚数字

heeng hahzahr
հինգ հազար

Ր

6,000 Armenian Numerals
Ⓕ 亚美尼亚数字

vehts hahzahr
վեց հազար

Ց

7,000 Armenian Numerals
Ⓕ 亚美尼亚数字

yote hahzahr
յոթ հազար

Ւ

8,000 Armenian Numerals
亚美尼亚数字

oote hahzahr
ութ հազար

Փ

9,000 Armenian Numerals
亚美尼亚数字

eene hahzahr
ինը հազար

Ք

10,000 Armenian Numerals
亚美尼亚数字

Overline placed above a letter multiplies its value by 10,000

Ա̅

1 (Ա) x 10,000 = 10,000
տաս հազար

100,000 Armenian Numerals
亚美尼亚数字

Overline placed above a letter multiplies its value by 10,000

Ճ̅

10 (Ճ) x 10,000 =100,000

1,000,000 Armenian Numerals Ⓕ 亚美尼亚数字	10,000,000 Armenian Numerals Ⓕ 亚美尼亚数字
Overline placed above a letter multiplies its value by 10,000	Overline placed above a letter multiplies its value by 10,000
Ճ̅ millihyone միլիոն	Ռ̅
100 (Ճ) x 10,000 = 1,000,000	1,000 (Ռ) x 10,000 = 10,000,000

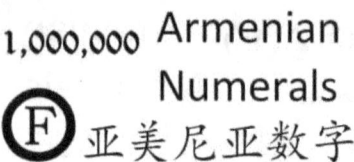

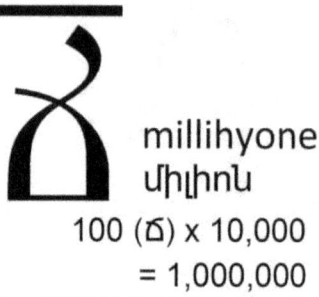

Georgian Numerals 乔治亚数字

Georgian belongs to the Kartvelian language family and has been written in various scripts through its history. The first Georgian script was in use in the 3rd century BC. Georgian is currently written in the Mkhedruli script which has 33 letters still in use. Letters of the Georgian alphabet are used to represent numbers which are written left to right in decreasing value order and added together. Either of two different alphabet letters can be used to write 400. Seven alphabet letters have no numerical value. The numbers from 1 to 19 have a base 10 or decimal structure. Numbers 11 to 19 are formed by prefixing numbers 1 to 9 with a shortened form of 10. A vigesimal (base 20) system is used for numbers 20 to 99; 40 is 2 x 20; 60 is 3 x 20 and 80 is 4 x 20. This is similar to the English usage of 3 score, 4 score etc. A base 10 structure operates within the base 20 groups. The Hundreds are formed by placing the numbers 2 through to 10 in front of the word for hundred.

0 Georgian Numerals
乔治亚数字

nuli
ნული

1 Georgian Numerals
乔治亚数字

erti
ერთი

ა

2 Georgian Numerals
乔治亚数字

ori
ორი

ბ

3 Georgian Numerals
乔治亚数字

sami
სამი

გ

4 Ⓘ
Georgian Numerals
乔治亚数字

otkhi
ოთხი

დ

5 Ⓘ
Georgian Numerals
乔治亚数字

khuti
ხუთი

ე

6 Ⓘ
Georgian Numerals
乔治亚数字

ekvsi
ექვსი

ვ

7 Ⓘ
Georgian Numerals
乔治亚数字

švidi
შვიდი

ზ

8 Ⓘ

Georgian Numerals

乔治亚数字

rva

რვა

ჱ

9 Ⓘ

Georgian Numerals

乔治亚数字

tskhra

ცხრა

თ

10 Ⓘ

Georgian Numerals

乔治亚数字

ati

ათი

ი

11 Ⓘ

Georgian Numerals

乔治亚数字

tertmet'i

თერთმეტი

ია

12 Georgian Numerals
Ⓘ 乔治亚数字

tormet'i
თორმეტი

იბ

13 Georgian Numerals
Ⓘ 乔治亚数字

tsamet'i
ცამეტი

იგ

14 Georgian Numerals
Ⓘ 乔治亚数字

totkhmet'i
თოთხმეტი

იდ

15 Georgian Numerals
Ⓘ 乔治亚数字

tkhutmet'i
თხუთმეტი

იე

16 Georgian Numerals

Ⓘ 乔治亚数字

tekvsmet'i

თექვსმეტი

ივ

17 Georgian Numerals

Ⓘ 乔治亚数字

čvidmet'i

ჩვიდმეტი

იზ

18 Georgian Numerals

Ⓘ 乔治亚数字

tvramet'i

თვრამეტი

იჱ

19 Georgian Numerals

Ⓘ 乔治亚数字

tskhramet'i

ცხრამეტი

ით

20 Georgian Numerals
Ⓘ 乔治亚数字

otsi
ოცი

კ

30 Georgian Numerals
Ⓘ 乔治亚数字

otsdaati
ოცდაათი

ლ

40 Georgian Numerals
Ⓘ 乔治亚数字

ormotsi
ორმოცი

მ

50 Georgian Numerals
Ⓘ 乔治亚数字

ormotsdaati
ორმოცდაათი

ნ

60 Georgian Numerals
Ⓘ 乔治亚数字

samotsi
სამოცი

ჲ

70 Georgian Numerals
Ⓘ 乔治亚数字

samotsdaati
სამოცდაათი

ო

80 Georgian Numerals
Ⓘ 乔治亚数字

otkhmotsi
ოთხმოცი

პ

90 Georgian Numerals
Ⓘ 乔治亚数字

otkhmotsdaati
ოთხმოცდაათი

ჟ

100 Georgian Numerals

Ⓘ 乔治亚数字

asi

ასი

რ

200 Georgian Numerals

Ⓘ 乔治亚数字

orasi

ორასი

ს

300 Georgian Numerals

Ⓘ 乔治亚数字

samasi

ამასი

ტ

400 Georgian Numerals

Ⓘ 乔治亚数字

otkhasi

ოთხასი

ჳ

400 Georgian Numerals

Ⓘ 乔治亚数字

otkhasi
ოთხასი

ჳ

500 Georgian Numerals

Ⓘ 乔治亚数字

khutasi
ხუთასი

ფ

600 Georgian Numerals

Ⓘ 乔治亚数字

ekvsasi
ექვსასი

ქ

700 Georgian Numerals

Ⓘ 乔治亚数字

švidasi
შვიდასი

ღ

800 Georgian Numerals
Ⓘ 乔治亚数字

rvaasi
რვაასი

ყ

900 Georgian Numerals
Ⓘ 乔治亚数字

tskhraasi
ცხრაასი

შ

1,000 Georgian Numerals
Ⓘ 乔治亚数字

atasi
ათასი

ჩ

2,000 Georgian Numerals
Ⓘ 乔治亚数字

ori atasi
ორი ათასი

ც

3,000 Georgian Numerals
(I) 乔治亚数字

sami atasi
სამი ათასი

ძ

4,000 Georgian Numerals
(I) 乔治亚数字

otkhi atasi
ოთხი ათასი

წ

5,000 Georgian Numerals
(I) 乔治亚数字

khuti atasi
ხუთი ათასი

ჭ

6,000 Georgian Numerals
(I) 乔治亚数字

ekvsi atasi
ექვსი ათასი

ხ

7,000 Georgian Numerals
乔治亚数字

švidi atasi
შვიდი ათასი

ჴ

8,000 Georgian Numerals
乔治亚数字

rva atasi
რვა ათასი

ჯ

9,000 Georgian Numerals
乔治亚数字

tskhra atasi
ცხრა ათასი

ჰ

10,000 Georgian Numerals
乔治亚数字

ati atasi
ათი ათასი

ჵ

Kharosthi Numerals 佉卢虱底数字

Kharosthi was a script that developed in the mid 3rd century BC in the ancient kingdom of Gandhara (modern day Afghanistan and Pakistan) and it was used in Gandharan Buddhism. Prakrit, Pali and Sanskrit were written in Kharosthi script which found its way to Central Asia where it was used along the Silk Road in Bactria, Sogdia and in the Kushan Empire. It died out in Afghanistan in the 3rd century but may have continued in use along the Silk Road until the 7th century. Kharosthi was mostly written right to left although some texts are written left to right which later became the standard in India. An analysis of the script indicates that it had its origins in the Aramaic alphabet which appears to have been extensively modified and adjusted to cope with local languages. The English scholar James Prinsep (1799-1840) used bilingual coins from the Indo-Greek kingdom of Gandhara to decipher the scrip. Greek was on the obverse and Pali on the reverse written in the Kharosthi script. He also deciphered the Brahmi script and his efforts enabled scholars to read the edicts of Ashoka, some of which were written in Kharosthi and Brahmi.

The Kharosthi numerals required repetition of symbols. For example, the symbol for 20 was written four times to represent 80. One was a single vertical stroke and two was at first written as two vertical strokes, and three as three vertical strokes. Eventually the strokes of two, and the strokes for three, were linked together almost exactly like the modern numerals for 2 and 3 in Arabic speaking countries. Four was a cross formed by two diagonal lines and eight was two such crosses. The symbol for 10 was doubled for 20. One could be placed before the symbol for hundred to indicate 100 and two before the symbol for 200 etc.

1 Kharosthi Numerals
Ω 佉卢 虱底数字

𐩀

2 Kharosthi Numerals
Ω 佉卢 虱底数字

𐩁

3 Kharosthi Numerals
Ω 佉卢 虱底数字

𐩂

4 Kharosthi Numerals
Ω 佉卢 虱底数字

𐩃

5 Kharosthi Numerals

佉卢 虱底数字

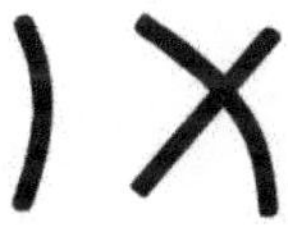

6 Kharosthi Numerals

佉卢 虱底数字

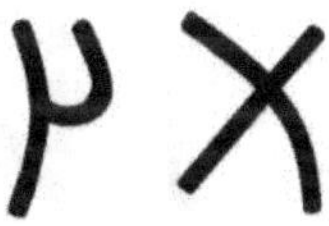

7 Kharosthi Numerals

佉卢 虱底数字

8 Kharosthi Numerals

佉卢 虱底数字

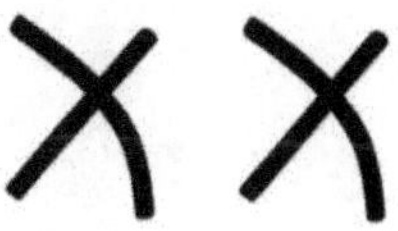

9 Kharosthi Numerals

Ω 佉卢虱底数字

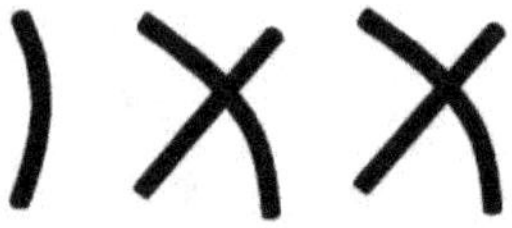

10 Kharosthi Numerals

Ω 佉卢虱底数字

11 Kharosthi Numerals

Ω 佉卢虱底数字

12 Kharosthi Numerals

Ω 佉卢虱底数字

13 Kharosthi Numerals

Ω 佉卢虱底数字

𐩄𐩂

14 Kharosthi Numerals

Ω 佉卢虱底数字

𐩄𐩃

15 Kharosthi Numerals

Ω 佉卢虱底数字

𐩄𐩃𐩀

16 Kharosthi Numerals

Ω 佉卢虱底数字

𐩄𐩃𐩁

17 Kharosthi Numerals

佉卢 虱底数字

𐩄𐩃𐩂

18 Kharosthi Numerals

佉卢 虱底数字

𐩄𐩃𐩃

19 Kharosthi Numerals

佉卢 虱底数字

𐩄𐩃𐩃𐩀

20 Kharosthi Numerals

佉卢 虱底数字

𐩅

30 Kharosthi Numerals
佉卢虱底数字

𐩅𐩄

40 Kharosthi Numerals
佉卢虱底数字

𐩅𐩅

50 Kharosthi Numerals
佉卢虱底数字

𐩅𐩅𐩄

60 Kharosthi Numerals
佉卢虱底数字

𐩅𐩅𐩅

70 Kharosthi Numerals

佉卢虱底数字

𐩅𐩅𐩅𐩄

80 Kharosthi Numerals

佉卢虱底数字

𐩅𐩅𐩅𐩅

90 Kharosthi Numerals

佉卢虱底数字

𐩅𐩅𐩅𐩅𐩄

100 Kharosthi Numerals

佉卢虱底数字

𐩀𐩆

1,000 Kharosthi Numerals

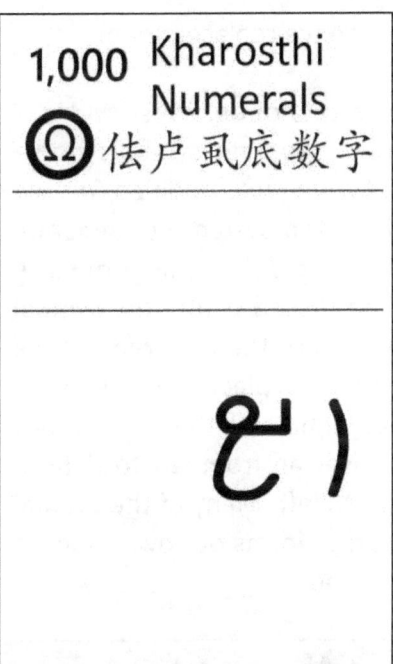

Ω 佉卢虱底数字

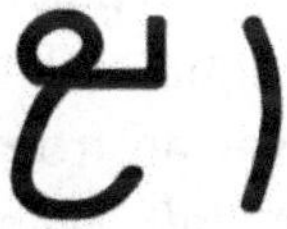

Brahmi Numerals 婆罗米数字

The Brahmi numerals developed in India around the middle of the 3rd century BC. They continued to develop and change further, becoming known as the Gupta numerals during the Gupta Dynasty which ruled the state of Magadha from the early 4th century to the late 6th century. The Gupta numerals continued to evolve and in the 7th century they were known as Devanagari numerals, the word Devanagari meaning "writing of the gods". Development continued beyond the 11th century until they reached the form currently used in writing Hindi and Sanskrit. There is no evidence that a place value system evolved in India before the late 6th century. The Gwalior inscription of 933 is the earliest undisputed Indian use of a place value system. However, Indian derived Khmer numerals in the Sambor inscriptions of 683 AD show the use of zero in the number 605 . Here zero is a dot and the Arabs used a dot for zero when they adopted Indian numerals in the 8th century. A circular or elliptical shape was already used in the symbols for 20, 80 and 90 in the Brahmi script and a circular zero may have replaced them.

Although Brahmi numerals are the ancestors of our Western Indo-Arabic numerals, they had neither nine separate symbols nor a symbol for zero. The horizontal stroke for one was doubled for two and tripled for three. Brahmi had a base-ten system but separate symbols for 10, 20, 30, up to 90 hindered the development of place values. On the other hand, two could be placed before the symbol for hundred to indicate 200 and three before the hundred symbol indicated 300. This was the beginning of a place value system. Brahmi numerals were written left to right and the adaptations made by the Arabs in the 8th century also ran from left to right in spite of Arabic script being written right to left. Many of the Brahmi numerals are clearly recognizable as earlier forms our own modern numerals, especially the numbers six to nine.

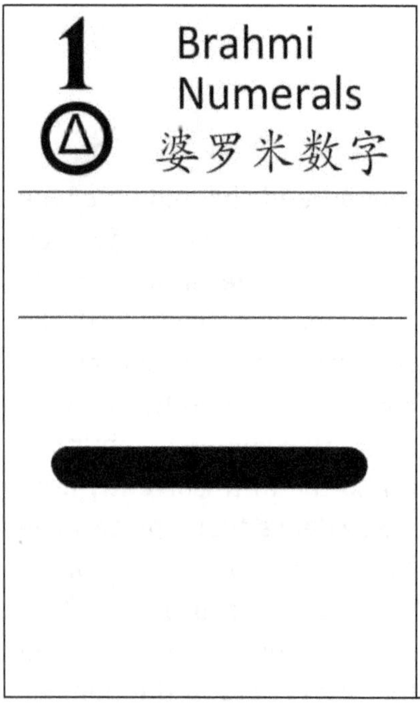

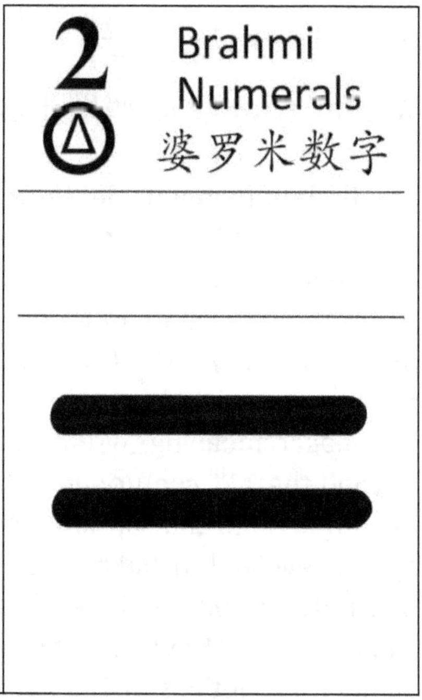

3

Brahmi
Numerals
婆罗米数字

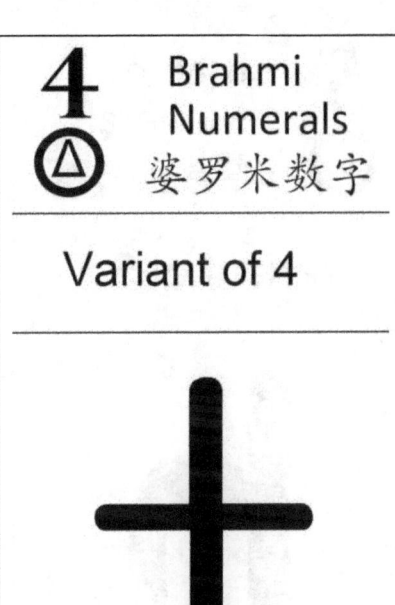

4

Brahmi
Numerals
婆罗米数字

Variant of 4

4

Brahmi
Numerals
婆罗米数字

Variant of 4

5

Brahmi
Numerals
婆罗米数字

Variant of 5

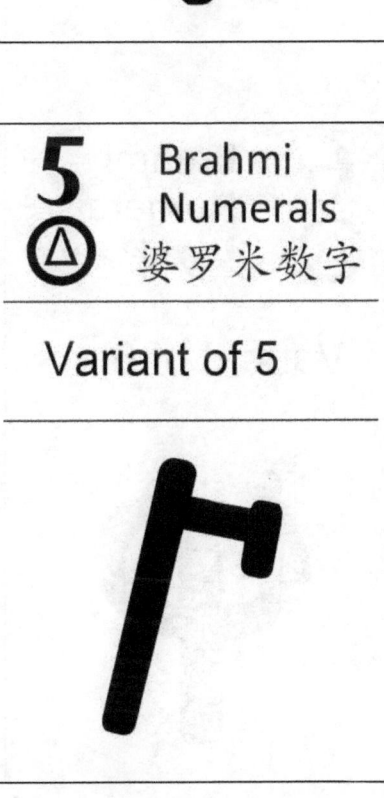

5

Brahmi Numerals
婆罗米数字

Variant of 5

6

Brahmi Numerals
婆罗米数字

Variant of 6

6

Brahmi Numerals
婆罗米数字

Variant of 6

7

Brahmi Numerals
婆罗米数字

Variant of 7

7 Brahmi Numerals 婆罗米数字

Variant of 7

7 Brahmi Numerals 婆罗米数字

Variant of 7

8 Brahmi Numerals 婆罗米数字

Variant of 8

8 Brahmi Numerals 婆罗米数字

Variant of 8

8

Brahmi Numerals
婆罗米数字

Variant of 8

9

Brahmi Numerals
婆罗米数字

Variant of 9

9

Brahmi Numerals
婆罗米数字

Variant of 9

10

Brahmi Numerals
婆罗米数字

11 Brahmi Numerals
婆罗米数字

12 Brahmi Numerals
婆罗米数字

13 Brahmi Numerals
婆罗米数字

14 Brahmi Numerals
婆罗米数字

Variant of 14

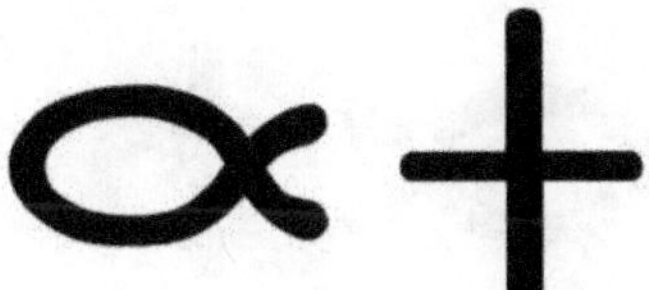

14 Brahmi Numerals 婆罗米数字

Variant of 14

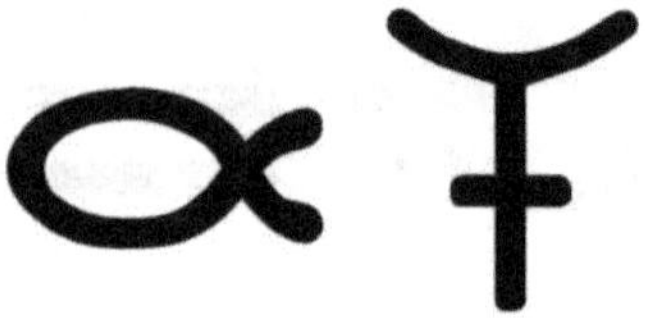

15 Brahmi Numerals 婆罗米数字

Variant of 15

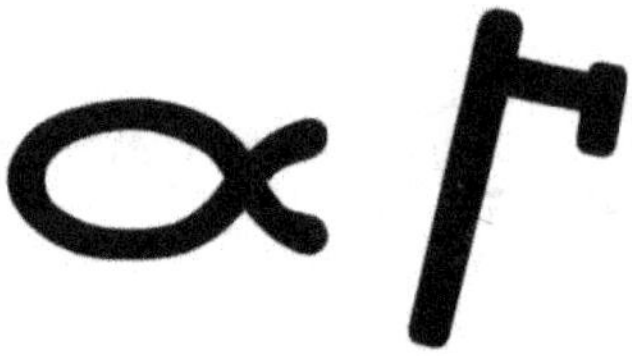

15 Brahmi Numerals 婆罗米数字

Variant of 15

16 Brahmi Numerals 婆罗米数字

Variant of 16

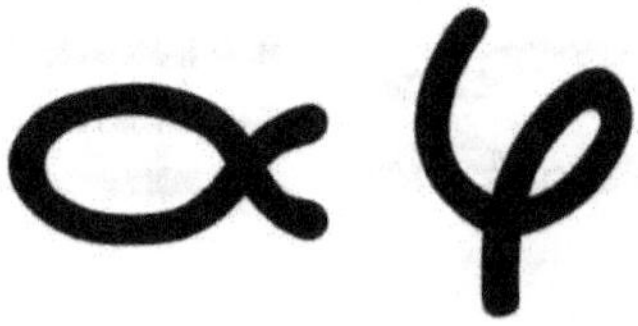

17 Brahmi Numerals
婆罗米数字

Variant of 17

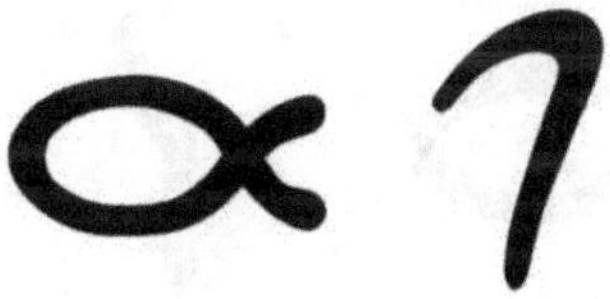

17 Brahmi Numerals
婆罗米数字

Variant of 17

18 Brahmi Numerals
婆罗米数字

Variant of 18

18 Brahmi Numerals
婆罗米数字

Variant of 18

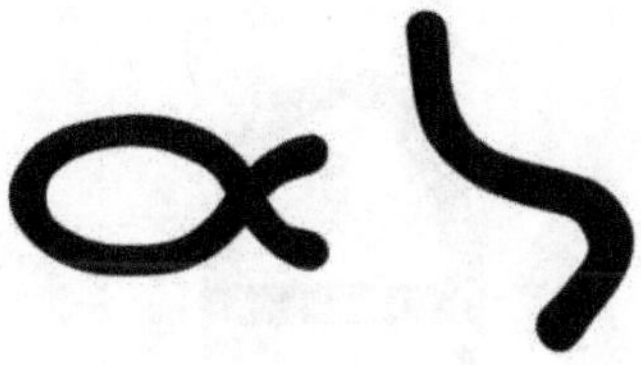

19 Brahmi Numerals 婆罗米数字

Variant of 19

19 Brahmi Numerals 婆罗米数字

Variant of 19

20 Brahmi Numerals 婆罗米数字

Variant of 20

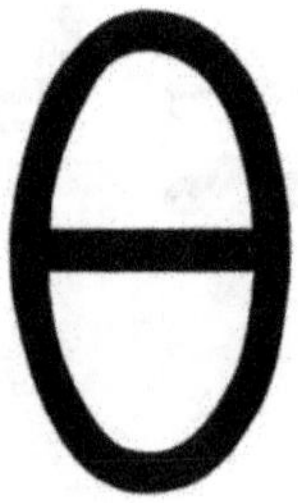

20 Brahmi Numerals 婆罗米数字

Variant of 20

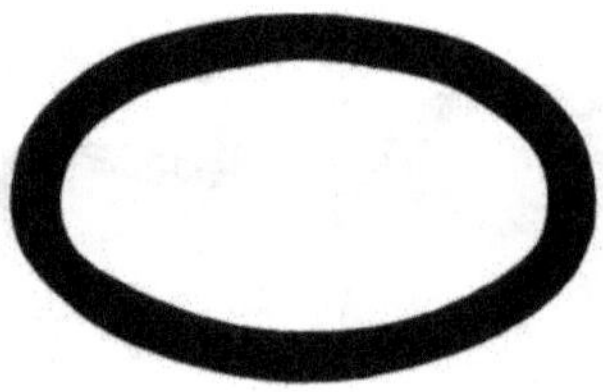

30 Brahmi Numerals

婆罗米数字

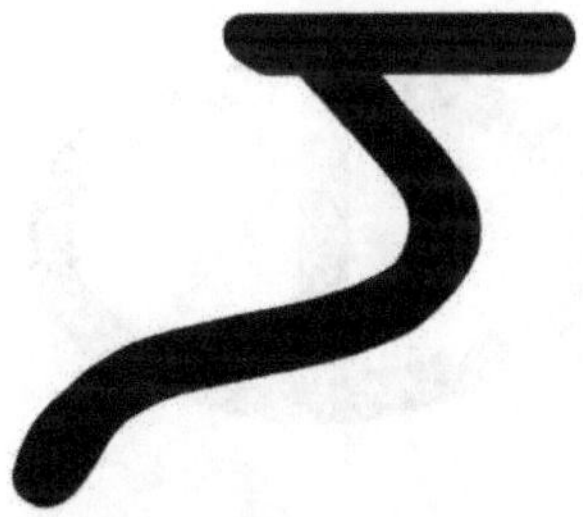

40 Brahmi Numerals

婆罗米数字

50 Brahmi Numerals

婆罗米数字

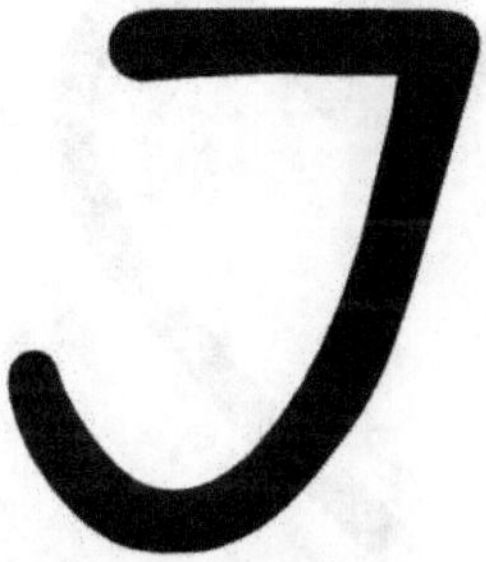

60 Brahmi Numerals

婆罗米数字

70 Brahmi Numerals

婆罗米数字

80 Brahmi Numerals

婆罗米数字

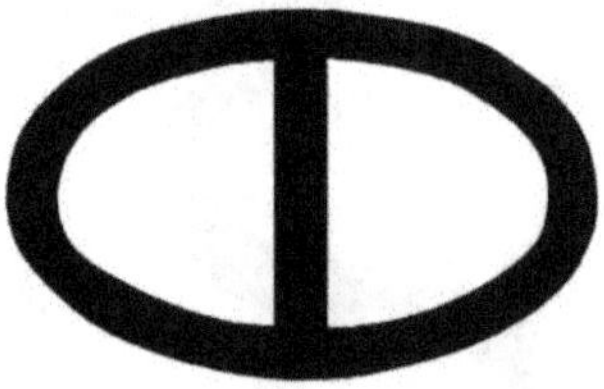

90 Brahmi Numerals

婆罗米数字

100 Brahmi Numerals

婆罗米数字

1,000 Brahmi Numerals 婆罗米数字

Sanskrit Numerals (Devanagari Script) 梵语数字（天城文）

Sanskrit, the cultural, literary and liturgical language of India, is usually written in the Devanagari script. Devanagari means "the writing of the gods" and it is also called Nagari. The script is written from left to right. There are 47 primary letters including 14 vowels and 33 consonants but there are no capital letters or small letters. Nagari evolved from the Brahmi script which originated in the mid third century BC. It had developed significantly by the 7th century and was fully developed by the 11th century. The numerals of the Brahmi script reached their final stage of development in Devanagari which has a base-ten decimal system with just nine symbols and a zero, and is thus equipped to record the largest numbers. Not only Sanskrit but also Hindi and other Indian languages are written in Devanagari script. Large Devanagari numerals are displayed below with their Sanskrit pronunciation.

0 Sanskrit Numerals 梵语数字

Śūnya

०

1 Sanskrit Numerals 梵语数字

Éka

१

2 Sanskrit Numerals 梵语数字

Dvi

२

3 Sanskrit Numerals 梵语数字

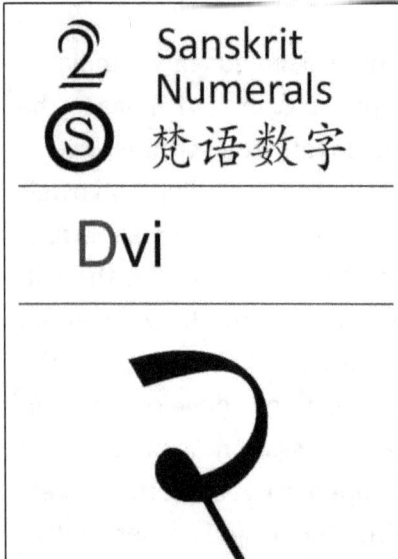

Trí

३

4 Sanskrit Numerals
梵语数字

Catúr

४

5 Sanskrit Numerals
梵语数字

Pañca

५

6 Sanskrit Numerals
梵语数字

Ṣáṣ

६

7 Sanskrit Numerals
梵语数字

Saptá

७

8 Sanskrit Numerals 梵语数字

Aṣṭạ

८

9 Sanskrit Numerals 梵语数字

Náva

९

10 Sanskrit Numerals 梵语数字

Dasa

१०

11 Sanskrit Numerals 梵语数字

Ékadasa

११

12 Sanskrit Numerals
梵语数字

Dvadasa

१२

13 Sanskrit Numerals
梵语数字

Trayodasa

१३

14 Sanskrit Numerals
梵语数字

Caturdasa

१४

15 Sanskrit Numerals
梵语数字

Pancadasa

१५

16 Sanskrit Numerals 梵语数字

Shash

१६

17 Sanskrit Numerals 梵语数字

Saptadasa

१७

18 Sanskrit Numerals 梵语数字

Aṣṭádasa

१८

19 Sanskrit Numerals 梵语数字

Návadasa

१९

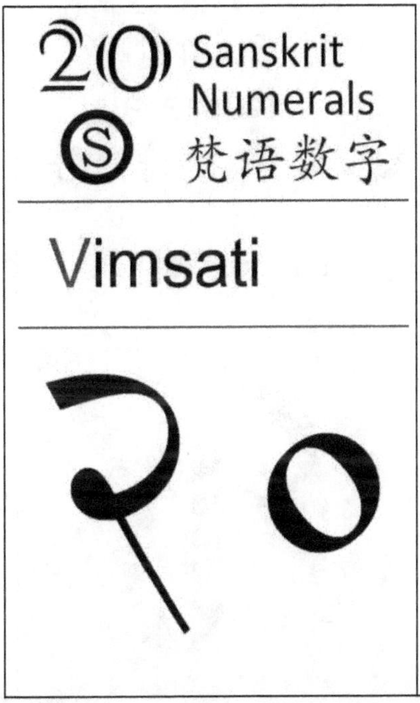

Gujarati Numerals 古吉拉特数字

Gujarati numerals are employed in the Gujarati language of Gujarat State in Western India. The Gujarati language is a derivative of Sanskrit and the Gujarati script is derived from the Devanagari script but has lost the long horizontal line running above the letters. Gujarati was the first language of Mahatma Gandhi (1869-1948) the leader of the Indian independence movement. Gujarat has a population of more than 60 million and borders Pakistan's Sindh Province. Several ancient cities of the Bronze Age Indus Valley civilization (3300-1300 BC) are located in Gujarat, including the port city of Lothal, India's first seaport.

0 Gujarati Numerals
古吉拉特数字

śūnya શૂન્ય

૦

1 Gujarati Numerals
古吉拉特数字

ek એક

૧

2 Gujarati Numerals
古吉拉特数字

be બે

૨

3 Gujarati Numerals
古吉拉特数字

tra ત્રણ

૩

4 Gujarati Numerals
古吉拉特数字

chār ચાર

૪

5 Gujarati Numerals
古吉拉特数字

pānch પાંચ

૫

6 Gujarati Numerals
古吉拉特数字

chha છ

૬

7 Gujarati Numerals
古吉拉特数字

sāt સાત

૭

8 Gujarati Numerals
古吉拉特数字

āth આઠ

૮

9 Gujarati Numerals
古吉拉特数字

nav નવ

૯

10 Gujarati Numerals
古吉拉特数字

das દસ

૧૦

11 Gujarati Numerals
古吉拉特数字

agiyār અગિયાર

૧૧

12 Gujarati Numerals
古吉拉特数字

bār બાર

૧૨

13 Gujarati Numerals
古吉拉特数字

tēr તેર

૧૩

14 Gujarati Numerals
古吉拉特数字

chaud ચૌદ

૧૪

15 Gujarati Numerals
古吉拉特数字

pamdar પંદર

૧૫

16 Gujarati Numerals
古吉拉特数字

sol સોળ

૧૬

17 Gujarati Numerals
古吉拉特数字

sattar સત્તર

૧૭

18 Gujarati Numerals
古吉拉特数字

adhār અઢાર

૧૮

19 Gujarati Numerals
古吉拉特数字

oganis ઓગણિસ

૧૯

20 Gujarati Numerals
Ⓖ 古吉拉特数字

vīs વીસ

૨૦

30 Gujarati Numerals
Ⓖ 古吉拉特数字

trīs ત્રીસ

૩૦

Gurmukhi Numerals 果鲁穆奇数字

The Gurmukhi numerals are used in the Gurmukhi script in India's Punjab. Gurmukhi is the script used by the second Sikh Guru, Guru Angad Dev Ji (1563-1605). Gurmukhi literally means "from the mouth of the Guru". The Gurmukhi syllabic alphabet has 35 letters and the script is written left to right. Two other scripts are used for the Punjabi language, the Nagari script used by Hundus and the Shahmukhi script used by Punjabi Muslims. Shahmukhi means "from the king's mouth". Punjabi is spoken by 130 million people in West Punjab, Pakistan and East Punjab, India. Punjabi became a separate language in the 11th century.

0 Gurmukhi Numerals
果鲁穆奇数字

sifar
ਸਿਫਰ

੦

1 Gurmukhi Numerals
果鲁穆奇数字

ikk
ਇੱਕ

੧

2 Gurmukhi Numerals
果鲁穆奇数字

do
ਦੋ

੨

3 Gurmukhi Numerals
果鲁穆奇数字

tinn
ਤਿੰਨ

੩

4 Gurmukhi Numerals
果鲁穆奇数字

chār
ਚਾਰ

੪

5 Gurmukhi Numerals
果鲁穆奇数字

pañj
ਪੰਜ

੫

6 Gurmukhi Numerals
果鲁穆奇数字

chhē
ਛੇ

੬

7 Gurmukhi Numerals
果鲁穆奇数字

satt
ਸੱਤ

੭

8 Gurmukhi Numerals
果鲁穆奇数字

atth
ਅੱਠ

੮

9 Gurmukhi Numerals
果鲁穆奇数字

naum
ਨੌਂ

੯

10 Gurmukhi Numerals
果鲁穆奇数字

das
ਦਸ

੧੦

11 Gurmukhi Numerals
果鲁穆奇数字

giārān
ਗਿਆਰਾ

੧੧

12 Gurmukhi Numerals
果鲁穆奇数字

baran
ਬਾਰਾ

੧੨

13 Gurmukhi Numerals
果鲁穆奇数字

teran
ਤੇਰਾਂ

੧੩

14 Gurmukhi Numerals
果鲁穆奇数字

chaudan
ਚੌਦਾਂ

੧੪

15 Gurmukhi Numerals
果鲁穆奇数字

pandran
ਪੰਦਰਾਂ

੧੫

16 Gurmukhi Numerals

果鲁穆奇数字

solan

ਸੋਲਾਂ

੧੬

17 Gurmukhi Numerals

果鲁穆奇数字

sataran

ਸਤਾਰਾਂ

੧੭

18 Gurmukhi Numerals

果鲁穆奇数字

ataran

ਅਠਾਰਾਂ

੧੮

19 Gurmukhi Numerals

果鲁穆奇数字

unni

ਉੱਨੀ

੧੯

20 Gurmukhi Numerals 果鲁穆奇数字	30 Gurmukhi Numerals 果鲁穆奇数字
vih ਵੀਹ	tīh ਤੀਹ
੨੦	੩੦

Bengali Numerals 孟加拉数字

Bengali numerals are used in the Bengali (Bangla) language which has more than 250 million native speakers making it the 6th most spoken native language in the world and the second most spoken native language in India. Bengali is the official language of Bangladesh and is an official language in the Indian states of West Bengal, Tripura, Assam and the Andaman and Nicobar Islands. The Bengali language evolved from Sanskrit and Maghadi Prakrit around 1000 AD and the alphabet and numerals are descendants of the Brahmi script. A base-ten decimal system with nine symbols and a zero are employed in Bengali. There is a large literature but the spoken and literary forms of the language vary somewhat in vocabulary and syntax.

0 Ⓝ

Bengali
Numerals
孟加拉数字

shunno শূন্য

০

1 Ⓝ

Bengali
Numerals
孟加拉数字

ak অক্

১

2 Ⓝ

Bengali
Numerals
孟加拉数字

dui ডুই

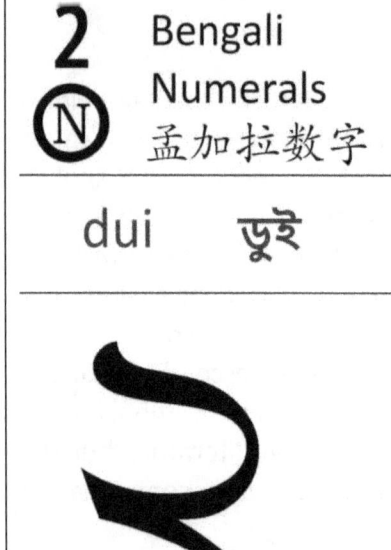

২

3 Ⓝ

Bengali
Numerals
孟加拉数字

tin টিন্

৩

4 Ⓝ Bengali Numerals 孟加拉数字

char চর্

৪

5 Ⓝ Bengali Numerals 孟加拉数字

pnach প্নচ্

৫

6 Ⓝ Bengali Numerals 孟加拉数字

soy সওয়্

৬

7 Ⓝ Bengali Numerals 孟加拉数字

sat সট্

৭

8 Ⓝ Bengali Numerals 孟加拉数字

aat আট্

৮

9 Ⓝ Bengali Numerals 孟加拉数字

noy নওয্

৯

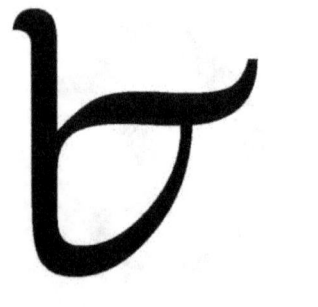

10 Ⓝ Bengali Numerals 孟加拉数字

dos ডওস্

১০

11 Ⓝ Bengali Numerals 孟加拉数字

agaro অগরও

১১

12 Bengali Numerals 孟加拉数字

baro বরও

১২

13 Bengali Numerals 孟加拉数字

tero টএরও

১৩

14 Bengali Numerals 孟加拉数字

choddo চওড্ডও

১৪

15 Bengali Numerals 孟加拉数字

ponero পওনএরও

১৫

16 Ⓝ Bengali Numerals 孟加拉数字

solo সওলও

১৬

17 Ⓝ Bengali Numerals 孟加拉数字

sotero সওটএরও

১৭

18 Ⓝ Bengali Numerals 孟加拉数字

atharo অতরও

১৮

19 Ⓝ Bengali Numerals 孟加拉数字

unish উনিশ্

১৯

20 Ⓝ Bengali Numerals 孟加拉数字	30 Ⓝ Bengali Numerals 孟加拉数字
bish বিশ্	trish ত্রিশ
২০	৩০

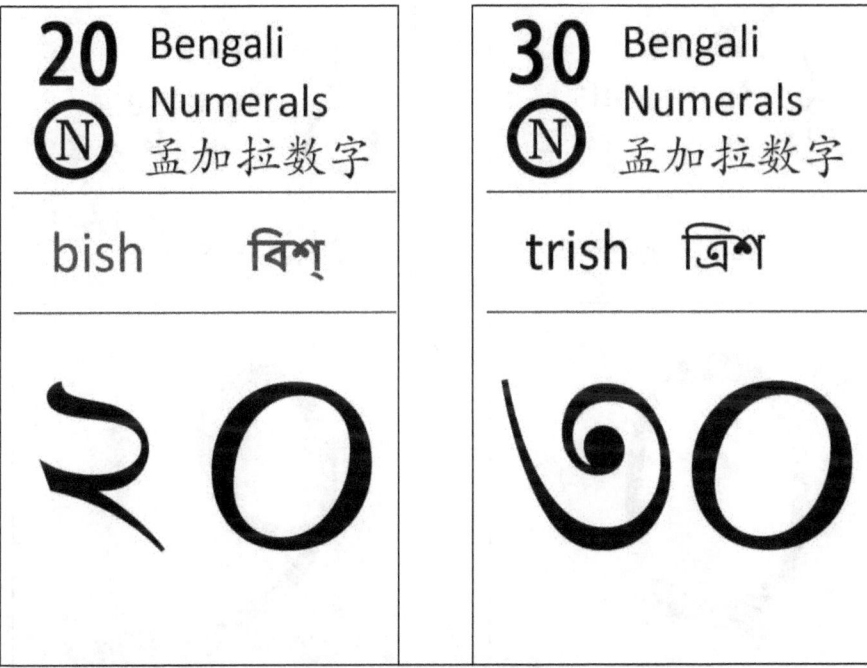

Oriya Numerals 奥里雅数字

Oriya (also called Odiya) is the official language of Odisha (Orissa) State in India's east. Odiya has a long literary history going back to the 7th century. It evolved from Magadhi Prakrit and is relatively pure with very few Arabic or Persian borrowings. Inscriptions in Odiya go back as far as the 1st century BC. It was barely distinguishable from Bengali until the 11th century. Odiya script evolved from Kalinga script which is a descendant of Brahmi. The curved letters result from writing on palm leaves which tend to crack or tear when straight lines are inscribed. The numerals are also mostly curved. Modern Odiya employs a base-ten numeral system with nine symbols and a zero.

0 Oriya Numerals 奥里雅数字

śunya ଶୂନ୍ୟ

୦

1 Oriya Numerals 奥里雅数字

eka ଏକ

୧

2 Oriya Numerals 奥里雅数字

dui ଦୁଇ

୨

3 Oriya Numerals 奥里雅数字

tini ତିନି

୩

4 Ⓤ Oriya Numerals 奥里雅数字

cāri ଚାରି

୪

5 Ⓤ Oriya Numerals 奥里雅数字

pānca ପାଞ୍ଚ

୫

6 Ⓤ Oriya Numerals 奥里雅数字

chaa ଛଅ

୬

7 Ⓤ Oriya Numerals 奥里雅数字

sāta ସାତ

୭

8 Ⓤ	Oriya Numerals 奥里雅数字
āṭha ଆଠ	
୮	

9 Ⓤ	Oriya Numerals 奥里雅数字
naa ନଅ	
୯	

10 Ⓤ	Oriya Numerals 奥里雅数字
daśa ନଅ	
୧୦	

11 Ⓤ	Oriya Numerals 奥里雅数字
egāra ଏଗାର	
୧୧	

12 Oriya Numerals

Ⓤ 奥里雅数字

bāra ବାର

୧୨

13 Oriya Numerals

Ⓤ 奥里雅数字

tera ତେର

୧୩

14 Oriya Numerals

Ⓤ 奥里雅数字

cauda ଚଉଦ

୧୪

15 Oriya Numerals

Ⓤ 奥里雅数字

pandara ପନ୍ଦର

୧୫

16 Oriya Numerals 奥里雅数字

sohala ଷୋହଳ

୧୬

17 Oriya Numerals 奥里雅数字

satara ସତର

୧୭

18 Oriya Numerals 奥里雅数字

athara ଅଠର

୧୮

19 Oriya Numerals 奥里雅数字

ūnāiśa ଊଣାଇଶ

୧୯

20 Ⓤ Oriya Numerals 奥里雅数字	30 Ⓤ Oriya Numerals 奥里雅数字
kodue କୋଡିଏ	tiriśi ତିରିଶି
୨୦	୩୦

Telugu Numerals 泰卢固数字

The Telugu numerals are used in the Telugu language which has more than 75 million speakers and is the official language of the states of Andhra Pradesh and Telangana located on the eastern side of peninsula India. Telugu is the most spoken member of the Dravidian family of languages found mainly in Southern India. Coins and inscriptions in Telugu date back as far as 500 BC. A script used jointly for Telugu and Kannada evolved from the Brahmi script but split into two scripts, one for each language, before the 15tn century. Development of a literary language began in the 11th century and the 16th century became a Golden Age of Telugu literature. The archaic style of Telugu writing underwent changes in the latter half of the 20th century reflecting the modern spoken language. Telugu is written left to right with a syllabic alphabet and has a base-10 system of numerals with nine symbols and a zero.

0 Telugu Numerals 泰卢固数字

sunna
సున్న

౦

1 Telugu Numerals 泰卢固数字

okati
ఒకటి

౧

2 Telugu Numerals 泰卢固数字

rendu
రెండు

౨

3 Telugu Numerals 泰卢固数字

mūdu
మూడు

౩

4 Telugu Numerals
泰卢固数字

nālugu
నాలుగు

౪

5 Telugu Numerals
泰卢固数字

ayidu
అయిదు

౫

6 Telugu Numerals
泰卢固数字

āru
ఆరు

౬

7 Telugu Numerals
泰卢固数字

ēdu
ఏడు

౭

8 Telugu Numerals
泰卢固数字

enimidi
ఎనిమిది

౮

9 Telugu Numerals
泰卢固数字

tommidi
తొమ్మిది

౯

10 Telugu Numerals
泰卢固数字

padi
పది

౧౦

11 Telugu Numerals
泰卢固数字

padakondu
పదకొండు

౧౧

12 Telugu Numerals 泰卢固数字

pannendu
పన్నెండు

౧౨

13 Telugu Numerals 泰卢固数字

padamūdu
పదమూడు

౧౩

14 Telugu Numerals 泰卢固数字

padhnālugu
పధ్నాలుగు

౧౪

15 Telugu Numerals 泰卢固数字

padunayidu
పదునయిదు

౧౫

16 Telugu Numerals
泰卢固数字

padahāru
పదహారు

౧౬

17 Telugu Numerals
泰卢固数字

padihēdu
పదిహేడు

౧౭

18 Telugu Numerals
泰卢固数字

padhdhenimidi
పద్ధెనిమిది

౧౮

19 Telugu Numerals
泰卢固数字

pandommidi
పందొమ్మిది

౧౯

20 Telugu Numerals 泰卢固数字	30 Telugu Numerals 泰卢固数字
iravai ఇరవై	muppai ముప్పై
౨౦	౩౦

Kannada Numerals 坎那达数字

The Kannada numerals are used in the Kannada language (also known as Kanarese) which is a member of the Dravidian family of languages and is spoken by more than 50 million people. Kannada is the official language of Karnataka State on the western side of peninsula India. In the 5th century the Kannada script evolved from the Kadamba script which was itself a descendant of the Brahmi script. By the year 1500 the Old Kannada script had split into the Kannada and Telugu scripts. Inscriptions in Kannada date back to the 5th century and inscribed coins to the 4th century; Kannada literature begins in the 9th century. Modern Kannada has a base-10 numeral system with nine symbols and a zero.

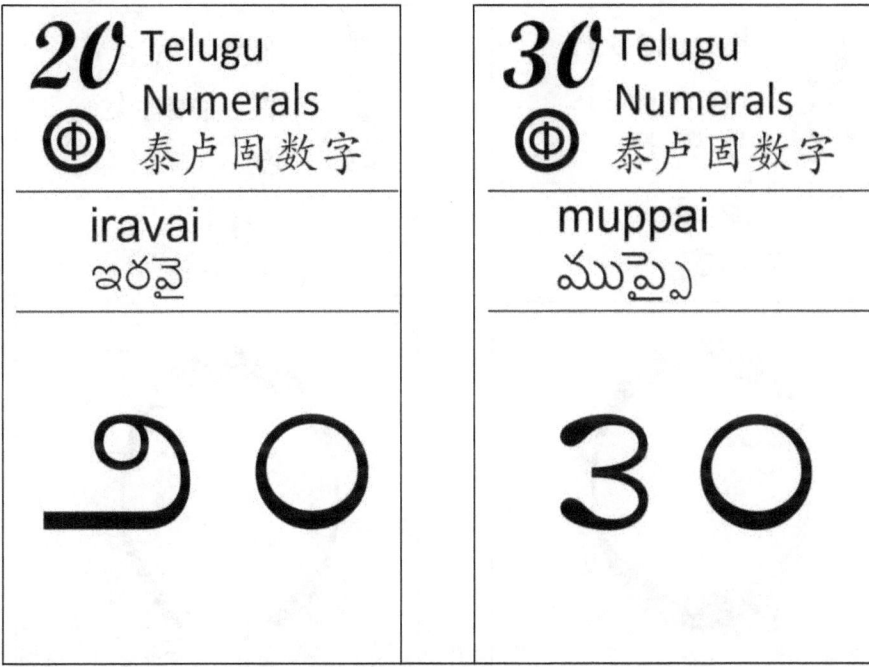

0

Kannada
Numerals
坎那达数字

sonnē
ಸೊನ್ನೆ

೦

1

Kannada
Numerals
坎那达数字

ondu
ಒಂದು

೧

2

Kannada
Numerals
坎那达数字

ēradu
ಎರಡು

೨

3

Kannada
Numerals
坎那达数字

mūru
ಮೂರು

೩

4 Ψ

Kannada
Numerals
坎那达数字

nālku
ನಾಲ್ಕು

5 Ψ

Kannada
Numerals
坎那达数字

aydu
ಅಯ್ದು

6 Ψ

Kannada
Numerals
坎那达数字

āru
ಆರು

7 Ψ

Kannada
Numerals
坎那达数字

ēlu
ಏಳು

8 Ψ

Kannada Numerals
坎那达数字

ēntu
ಎಂಟು

೮

9 Ψ

Kannada Numerals
坎那达数字

ombattu
ಒಂಬತ್ತು

೯

10 Ψ

Kannada Numerals
坎那达数字

hattu
ಹತ್ತು

೧೦

11 Ψ

Kannada Numerals
坎那达数字

hannondu
ಹನ್ನೊಂದು

೧೧

12 Kannada Numerals
Ⓟ 坎那达数字

hannēradu
ಹನ್ನೆರಡು

೧೨

13 Kannada Numerals
Ⓟ 坎那达数字

hadimūru
ಹದಿಮೂರು

೧೩

14 Kannada Numerals
Ⓟ 坎那达数字

hadinālku
ಹದಿನಾಲ್ಕು

೧೪

15 Kannada Numerals
Ⓟ 坎那达数字

hadinaidu
ಹದಿನೈದು

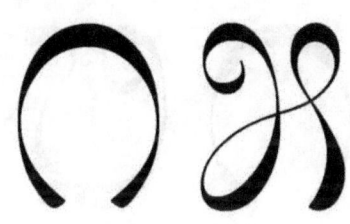

೧೫

16 Ψ

Kannada Numerals
坎那达数字

hadināaru
ಹದಿನಾರು

೧೬

17 Ψ

Kannada Numerals
坎那达数字

hadinēlu
ಹದಿನೇಳು

೧೭

18 Ψ

Kannada Numerals
坎那达数字

hadinēntu
ಹದಿನೆಂಟು

೧೮

19 Ψ

Kannada Numerals
坎那达数字

hattombattu
ಹತ್ತೊಂಬತ್ತು

೧೯

20 Kannada Numerals 坎那达数字	30 Kannada Numerals 坎那达数字
ippattu ಇಪ್ಪತ್ತು	mūvattu ಮೂವತ್ತು
೨೦	೩೦

Tamil Numerals 泰米尔数字

Tamil numerals are used in the Tamil language which is a member of the Dravidian language family. Tamil is spoken by about 70 million people in the Indian state of Tamil Nadu and in Sri Lanka, Singapore and Malaysia as well as in other countries such as South Africa, Mauritius and Fiji. Tamil inscriptions in the Brahmi script go back to the 3rd century BC. The modern Tamil script appeared in the 7th century and was based on the Grantha script, a descendant of Brahmi. There is a great variety of classical Tamil literature which is of high quality and there is continuity to the modern period. Tamil has a syllabic script and is written left to right. The numerals traditionally had no zero. There are symbols for one to nine and separate symbols for 10, 100 and 1,000. For eleven the symbol for 10 is followed by the symbol for one and this system continues to 19. Then twenty uses the symbol for two followed by the symbol for 10 and this continues up to 90.

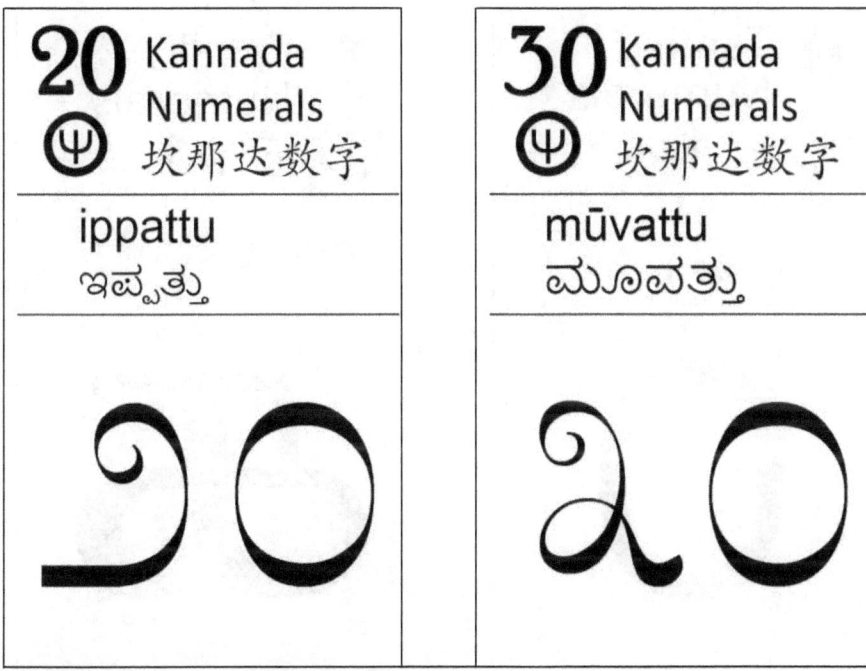

0 Tamil Numerals
泰米尔数字

Chuliyam
(pūjjiyam)

௦

1 Tamil Numerals
泰米尔数字

ondru

௧

2 Tamil Numerals
泰米尔数字

irantu

௨

3 Tamil Numerals
泰米尔数字

moondru

௩

4 ⓣ Tamil Numerals 泰米尔数字

nānku

௪

5 ⓣ Tamil Numerals 泰米尔数字

aindhu

௫

6 ⓣ Tamil Numerals 泰米尔数字

āru

௬

7 ⓣ Tamil Numerals 泰米尔数字

ezhu

௭

8 Tamil Numerals
Ⓣ 泰米尔数字

ettu

௮

9 Tamil Numerals
Ⓣ 泰米尔数字

onpadhu

௯

10 Tamil Numerals
Ⓣ 泰米尔数字

paththu

௰

11 Tamil Numerals
Ⓣ 泰米尔数字

padhinondru

௰௧

12 Tamil Numerals

Ⓣ 泰米尔数字

pannirentu

௰௨

13 Tamil Numerals

Ⓣ 泰米尔数字

padhinmoondru

௰௩

14 Tamil Numerals

Ⓣ 泰米尔数字

padhinānku

௰௪

15 Tamil Numerals

Ⓣ 泰米尔数字

padhinaindhu

௰௫

16 Tamil Numerals
Ⓣ 泰米尔数字

padhināru

௰௬

17 Tamil Numerals
Ⓣ 泰米尔数字

padhinezhu

௰௭

18 Tamil Numerals
Ⓣ 泰米尔数字

padhinettu

௰௮

19 Tamil Numerals
Ⓣ 泰米尔数字

paththonpadhu

௰௯

20 Tamil Numerals

Ⓣ 泰米尔数字

irupatu

௨௰

30 Tamil Numerals

Ⓣ 泰米尔数字

muppatu

௩௰

40 Tamil Numerals

Ⓣ 泰米尔数字

nārpatu

௪௰

50 Tamil Numerals

Ⓣ 泰米尔数字

aimpatu

௫௰

60 Tamil Numerals
Ⓣ 泰米尔数字

arupatu

௬௰

70 Tamil Numerals
Ⓣ 泰米尔数字

elupatu

௭௰

80 Tamil Numerals
Ⓣ 泰米尔数字

enpatu

௮௰

90 Tamil Numerals
Ⓣ 泰米尔数字

tonnūru

௯௰

100 Tamil Numerals 泰米尔数字

nūru 100

௱

1,000 Tamil Numerals 泰米尔数字

āyiram 1,000

௲

100,000 Tamil Numerals 泰米尔数字

nūraiyiram latcam 100,000

௱௲

one million Tamil Numerals 泰米尔数字

meiyyiram pattu latcam 1,000,000

௲௲

one trillion

Tamil Numerals

Ⓣ 泰米尔数字

tollun nikarputam

1,000,000,000,000

௲௲௲

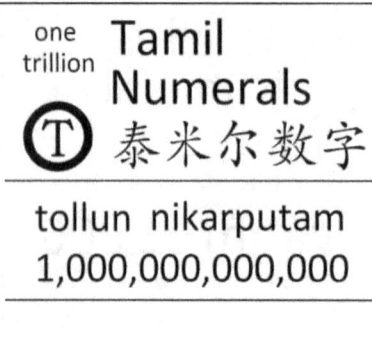

Malayalam Numerals 马拉雅拉姆数字

Malayalam numerals are used in the Malayalam language which is a member of the Dravidian language family. About 40 million people in the southwest of India speak Malayalam which is the official language of Kerala State. The earliest writing in Malayalam was in the Vatteluttu script, a descendent of Brahmi. Malayalam may have separated from Tamil in the 8th century and in the 8th or 9th century the Grantha script, another descendant of Brahmi, was introduced and adapted. By the early 13th century a standardized Malayalam alphabet had emerged. The Malayalam script and numerals are written left to right. There was traditionally no zero; there are symbols for one to nine and separate symbols for 10, 100 and 1,000. For eleven, the symbol for 10 is followed by the symbol for one and this system continues to 19. Then twenty uses the symbol for two followed by the symbol for 10 and this continues up to 90.

0 Malayalam Numerals

马拉雅拉姆数字

poojyam
പൂജ്യം

൦

1 Malayalam Numerals

马拉雅拉姆数字

onnu
ഒന്ന്

൧

2 Malayalam Numerals

马拉雅拉姆数字

randu
രണ്ട്

൨

3 Malayalam Numerals

马拉雅拉姆数字

moonnu
മൂന്ന്

൩

4 Malayalam Numerals
马拉雅拉姆数字

naalu
നാല്

൪

5 Malayalam Numerals
马拉雅拉姆数字

anchu
അഞ്ച്

൫

6 Malayalam Numerals
马拉雅拉姆数字

aaru
അറ്

൬

7 Malayalam Numerals
马拉雅拉姆数字

eezhu
ഏഴ്

൭

8 Malayalam Numerals
马拉雅拉姆数字

ettu
എട്ട്

൮

9 Malayalam Numerals
马拉雅拉姆数字

onpathu
ഒന്പത്

൯

10 Malayalam Numerals
马拉雅拉姆数字

pathu
പത്ത്

൰

11 Malayalam Numerals
马拉雅拉姆数字

pathinonnuh
പതിനൊന്ന്

൰൧

12 Malayalam Numerals
马拉雅拉姆数字

panthranduh
പന്ത്രണ്ട്

൧൨

13 Malayalam Numerals
马拉雅拉姆数字

pathimoonnuh
പതിമൂന്നു

൧൩

14 Malayalam Numerals
马拉雅拉姆数字

pathinaaluh
പതിനാല്

൧൪

15 Malayalam Numerals
马拉雅拉姆数字

pathinanchuh
പതിനഞ്ച്

൧൫

16 Malayalam Numerals

Σ 马拉雅拉姆数字

pathinaaruh

പതിനാറ്

൧൬

17 Malayalam Numerals

Σ 马拉雅拉姆数字

pathinezhuh

പതിനേഴ്

൧൭

18 Malayalam Numerals

Σ 马拉雅拉姆数字

pathinettuh

പതിനെട്ട്

൧൮

19 Malayalam Numerals

Σ 马拉雅拉姆数字

pathombathuh

പത്തൊമ്പതു

൧൯

20 Malayalam Numerals

Σ 马拉雅拉姆数字

irupathuh

ഇരുപത്

൨൰

30 Malayalam Numerals

Σ 马拉雅拉姆数字

muppathuh

മുപ്പത്

൩൰

40 Malayalam Numerals

Σ 马拉雅拉姆数字

nalpathuh

നാല്പത്

൪൰

50 Malayalam Numerals

Σ 马拉雅拉姆数字

anpathuh

അന്പത്

൫൰

60 Malayalam Numerals
Ⓢ 马拉雅拉姆数字

arupathuh
അറുപത്

൬൰

---

70 Malayalam Numerals
Ⓢ 马拉雅拉姆数字

ezhupathuh
എഴുപത്

൭൰

---

80 Malayalam Numerals
Ⓢ 马拉雅拉姆数字

enpathuh
എണ്പത്

൮൰

---

90 Malayalam Numerals
Ⓢ 马拉雅拉姆数字

thonnooru
തൊണ്ണൂറ്

൯൰

100 Malayalam Numerals
马拉雅拉姆数字

nooru
നൂറ്

൱

1,000 Malayalam Numerals
马拉雅拉姆数字

aayiram
ആയിരം

൲

1,000,000 Malayalam Numerals
马拉雅拉姆数字

pathu-laksham
പത്തുലക്ഷം

൲൲

Sinhala Numerals 斯里兰卡僧伽罗数字

The archaic Sinhala numerals are no longer used and have been replaced by Western Indo-Arabic numerals. Sinhala numerals did not have a zero and there were separate symbols for 10, 20 and 30 up to 100 and also for 1,000. The Sinhala language, spoken by nearly 20 million people, is a member of the Indo-Aryan language family and the Sinhala script is a descendant of the Brahmi script. Sinhala is written left to right and its letters are more circular or round-shaped than those of any other Indic language. Curved lines had to be used when writing on palm leaves with a stylus to prevent the leaves splitting or tearing. The earliest inscriptions are from the 3rd century BC and the earliest surviving literature is from the 9th century.

0 Sri Lankan Sinhala Numerals
斯里兰卡僧伽罗数字

binduva බින්දුව

---

1 Sri Lankan Sinhala Numerals
斯里兰卡僧伽罗数字

eka එක

2 Sri Lankan Sinhala Numerals

斯里兰卡僧伽罗数字

deka දෙක

3 Sri Lankan Sinhala Numerals

斯里兰卡僧伽罗数字

thuna තුන

4 Sri Lankan Sinhala Numerals

斯里兰卡僧伽罗数字

hathara හතර

5 Sri Lankan Sinhala Numerals

斯里兰卡僧伽罗数字

paha පහ

6 Ⓩ Sri Lankan Sinhala Numerals

斯里兰卡僧伽罗数字

haya හය

7 Ⓩ Sri Lankan Sinhala Numerals

斯里兰卡僧伽罗数字

hatha හත

8 Ⓩ Sri Lankan Sinhala Numerals

斯里兰卡僧伽罗数字

ata අට

9 Ⓩ Sri Lankan Sinhala Numerals

斯里兰卡僧伽罗数字

navaya නවය
(namaya) (නමය)

10 Ⓩ Sri Lankan Sinhala Numerals
斯里兰卡僧伽罗数字

dahaya දහය

𑇪

11 Ⓩ Sri Lankan Sinhala Numerals
斯里兰卡僧伽罗数字

ekolaha
එකොළහ

𑇪 𑇡

12 Ⓩ Sri Lankan Sinhala Numerals
斯里兰卡僧伽罗数字

dolaha
දොළහ

𑇪 𑇢

13 Ⓩ Sri Lankan Sinhala Numerals
斯里兰卡僧伽罗数字

dahathuna
දහතුන

𑇪 𑇣

14 Sri Lankan Sinhala Numerals

斯里兰卡僧伽罗数字

dahahathara
දහහතර
(dahathara)
(දාහතර)

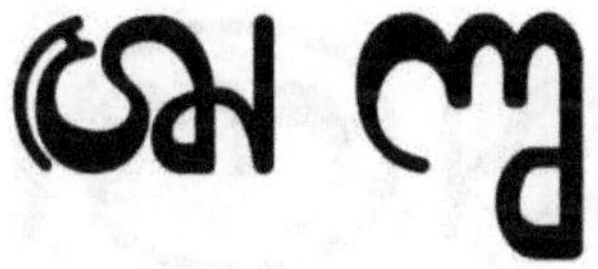

15 Sri Lankan Sinhala Numerals

斯里兰卡僧伽罗数字

pahalova
පහළොව

16 Sri Lankan Sinhala Numerals

斯里兰卡僧伽罗数字

dahasaya දහසය
(dasaya) (දාසය)

17 Sri Lankan Sinhala Numerals

斯里兰卡僧伽罗数字

dahahatha (dahata)
දහහත (දාහත)

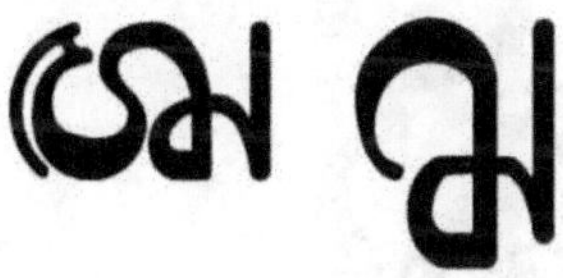

18 Ⓩ Sri Lankan Sinhala Numerals

斯里兰卡僧伽罗数字

dahaata දහඅට

19 Ⓩ Sri Lankan Sinhala Numerals

斯里兰卡僧伽罗数字

dahanavaya
දහනවය
(dahanamaya)
(දහනමය)

20 Ⓩ Sri Lankan Sinhala Numerals

斯里兰卡僧伽罗数字

visa විස්ස

30 Ⓩ Sri Lankan Sinhala Numerals

斯里兰卡僧伽罗数字

thiha තිහ

40 Sri Lankan Sinhala Numerals

斯里兰卡僧伽罗数字

hathaliha හතලිහ

𑇭

50 Sri Lankan Sinhala Numerals

斯里兰卡僧伽罗数字

panaha පනහ

𑇮

60 Sri Lankan Sinhala Numerals

斯里兰卡僧伽罗数字

hata හැට

𑇯

70 Sri Lankan Sinhala Numerals

斯里兰卡僧伽罗数字

haththawa හැත්තෑව

𑇰

80 Ⓩ Sri Lankan Sinhala Numerals

斯里兰卡僧伽罗数字

asuwa අසූව

𑇱

90 Ⓩ Sri Lankan Sinhala Numerals

斯里兰卡僧伽罗数字

anuwa අනූව

𑇲

100 Ⓩ Sri Lankan Sinhala Numerals

斯里兰卡僧伽罗数字

siiya සීය

𑇳

1,000 Ⓩ Sri Lankan Sinhala Numerals

斯里兰卡僧伽罗数字

daaha දාහ

𑇴

Javanese Numerals 爪哇数字

Javanese is an Austronesian language belonging to the Malayo-polynesian family of languages. It is spoken by more than 100 million people principally in Central and East Java. In the 4th century Javanese was written in the Pallava script derived from India, which by the 10th century had evolved into the Kawi script. By the 17th century the Javanese script had developed into its current form. Javanese has 53 letters and is written left to right. It has three different registers or styles, each employing its own vocabulary. The lower style (*ngoko*) is a humble form while *karma* (pronounced "kromo") is an honorific form and *karma inggil* is extremely honorific. *Madya* is an intermediate form between *ngoko* and *karma*. Old Javanese and Sanskrit words are known as *kawi* and numerals have *ngoko, karma* and *kawi* forms. Most *kawi* forms of the numerals are very close to Sanskrit.

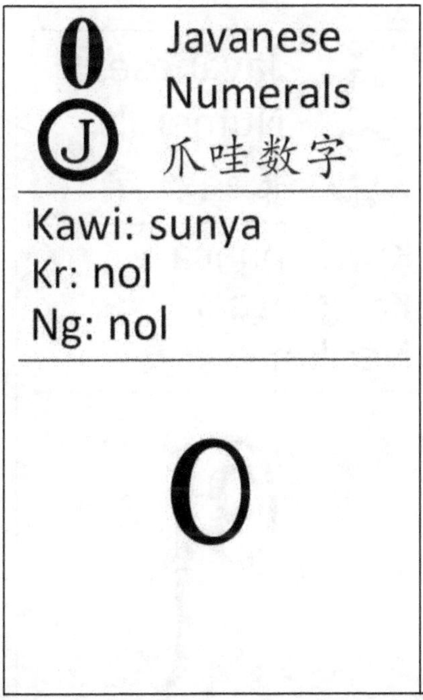
0
J
Javanese
Numerals
爪哇数字
Kawi: sunya
Kr: nol
Ng: nol

1
J
Javanese
Numerals
爪哇数字
Kawi: eka
Kr: setunggal
Ng: siji

2 Ⓙ

Javanese Numerals

爪哇数字

Kawi: dwi
Kr: kalih
Ng: loro

꧒

3 Ⓙ

Javanese Numerals

爪哇数字

Kawi: tri
Kr: tiga
Ng: telu

꧓

4 Ⓙ

Javanese Numerals

爪哇数字

Kawi: catur
Kr: sekawan
Ng: papat

꧔

5 Ⓙ

Javanese Numerals

爪哇数字

Kawi: panca
Kr: gangsal
Ng: lima

꧕

6 Javanese Numerals
Ⓙ 爪哇数字

Kawi: sad
Kr: enem
Ng: enem

7 Javanese Numerals
Ⓙ 爪哇数字

Kawi: sapta
Kr: pitu
Ng: pitu

8 Javanese Numerals
Ⓙ 爪哇数字

Kawi: asta
Kr: wolu
Ng: wolu

9 Javanese Numerals
Ⓙ 爪哇数字

Kawi: nawa
Kr: sanga
Ng: sanga

10

Ⓙ

Javanese Numerals
爪哇数字

Kawi: dasa
Kr: sedasa
Ng: sepuluh

꧑꧐

11

Ⓙ

Javanese Numerals
爪哇数字

Kawi: ekadasa
Kr: setunggal welas
Ng: sewelas

꧑꧑

12

Ⓙ

Javanese Numerals
爪哇数字

Kawi: dwidasa
Kr: kalih welas
Ng: rolas

꧑꧒

13

Ⓙ

Javanese Numerals
爪哇数字

Kawi: tridasa
Kr: tiga welas
Ng: telulas

꧑꧓

14 Ⓙ Javanese Numerals 爪哇数字

Kawi: caturdasa
Kr: sekawan welas
Ng: patbelas

꧑꧔

15 Ⓙ Javanese Numerals 爪哇数字

Kawi: pancadasa
Kr: gangsal welas
Ng: limalas

꧑꧕

16 Ⓙ Javanese Numerals 爪哇数字

Kawi: sadasa
Kr: enem welas
Ng: nembelas

꧑꧖

17 Ⓙ Javanese Numerals 爪哇数字

Kawi: saptadasa
Kr: pitu welas
Ng: pitulas

꧑꧗

18 ⓙ Javanese Numerals 爪哇数字

Kawi: astadasa
Kr: wolu welas
Ng: wolulas

꧑꧘

19 ⓙ Javanese Numerals 爪哇数字

Kawi: nawadasa
Kr: sanga welas
Ng: sangalas

꧑꧙

20 ⓙ Javanese Numerals 爪哇数字

Kawi: wingsati
Kr: kalih dasa
Ng: rong puluh

꧒꧐

21 ⓙ Javanese Numerals 爪哇数字

Kawi: eka-dwidasa
Kr: setunggal likur
Ng: selikur

꧒꧑

22 Ⓙ

Javanese Numerals
爪哇数字

Kawi: dwi-dwidasa
Kr: kalih likur
Ng: rolikur/lolikur

꧒꧒

23 Ⓙ

Javanese Numerals
爪哇数字

Kawi: tri-dwidasa
Kr: tiga likur
Ng: telulikur

꧒꧓

24 Ⓙ

Javanese Numerals
爪哇数字

Kawi: catur-dwidasa
Kr: sekawan likur
Ng: patlikur

꧒꧔

25 Ⓙ

Javanese Numerals
爪哇数字

Kawi: panca-dwidasa
Kr: selangkung
Ng: selawé

꧒꧕

26 Ⓙ Javanese Numerals 爪哇数字

Kawi: sad-dwidasa
Kr:
Ng: nemlikur

꧒꧖

27 Ⓙ Javanese Numerals 爪哇数字

Kawi: sapta-dwidasa
Kr:
Ng: pitulikur

꧒꧗

28 Ⓙ Javanese Numerals 爪哇数字

Kawi: asta-dwidasa
Kr:
Ng: wolulikur

꧒꧘

29 Ⓙ Javanese Numerals 爪哇数字

Kawi: nawa-dwidasa
Kr:
Ng: sangalikur

꧒꧙

30 Javanese Numerals

Ⓙ 爪哇数字

Kawi: trinisat

Kr: tigang dasa

Ng: telung puluh

꧓꧐

40 Javanese Numerals

Ⓙ 爪哇数字

Kawi: catrawingsat

Kr: patang dasa

Ng: patang puluh

꧔꧐

Burmese Numerals 缅甸数字

The Burmese script was adapted from the Old Mon or Pyu scripts which evolved from the South Indian Kadamba or Pallava scripts both of which are in turn derived from the Brahmi script. The history of writing in Myanmar goes back more than 1,000 years. Burmese was originally written in a squared format but a cursive form developed in the 17th century when writing on palm leaves became popular. Making straight lines with a stylus could rip or tear the leaves and cursive strokes avoided this. Burmese has 33 letters and is written left to right. It is a member of the Sino-Tibetan language family and is spoken as a first language by more than 35 million people mostly of Bamar or Burman origin and as a second language by ethnic minorities who number more than 10 million. Burmese is a tonal language. There is no standard system for writing Burmese language or numerals with Roman letters.

0 Ⓨ Myanmar Burmese Numerals 缅甸数字

su.nya. θòuɴɲa̰ သုည

၀

1 Ⓨ Myanmar Burmese Numerals 缅甸数字

tiq
tit
tiʔ တစ်

၁

2 Ⓨ Myanmar Burmese Numerals 缅甸数字

hniq
hnit
n̥iʔ နှစ်

၂

3 Ⓨ Myanmar Burmese Numerals 缅甸数字

thòun
thone
θóuɴ သုံး

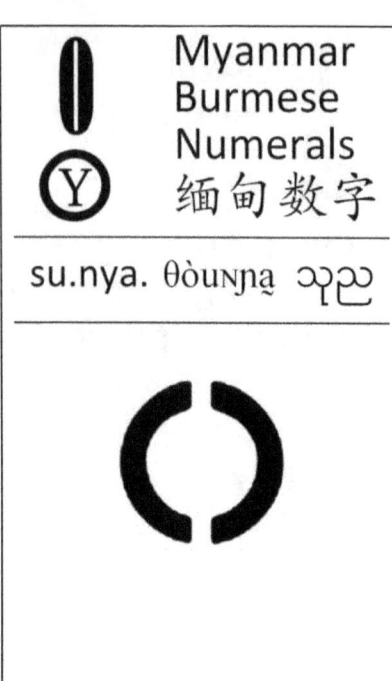

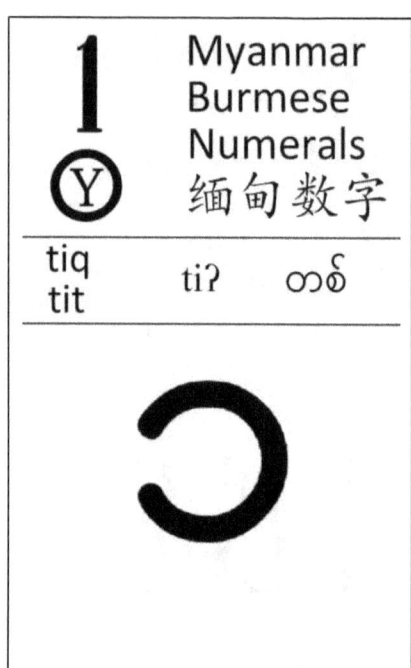

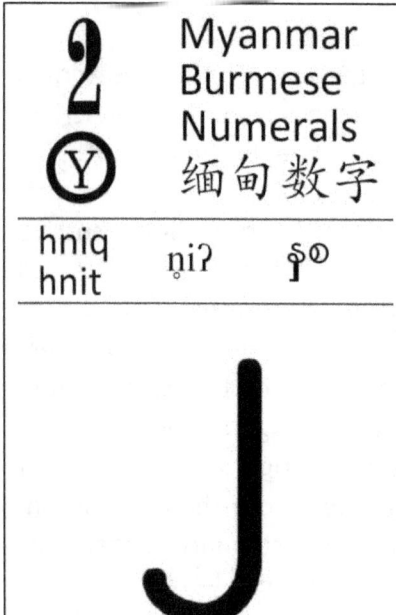

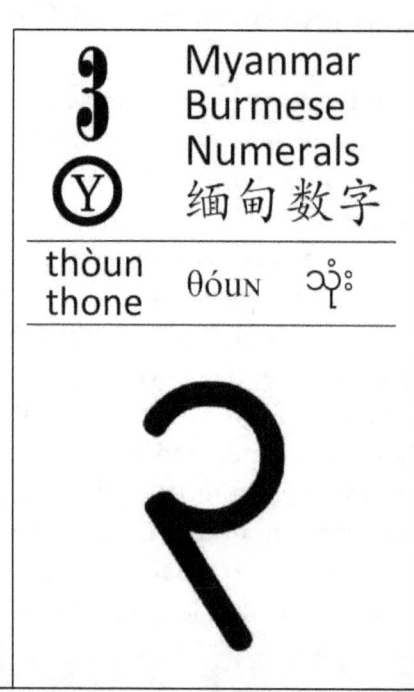

၃

4 Ⓨ

Myanmar
Burmese
Numerals
缅甸数字

lè
lay

lé

လေး

၄

5 Ⓨ

Myanmar
Burmese
Numerals
缅甸数字

ngà
nga

ŋá
ngarr

ငါး

၅

6 Ⓨ

Myanmar
Burmese
Numerals
缅甸数字

c'auq
chout

tɕʰauʔ

ခြောက်

၆

7 Ⓨ

Myanmar
Burmese
Numerals
缅甸数字

k'u-hniq
khun nhit

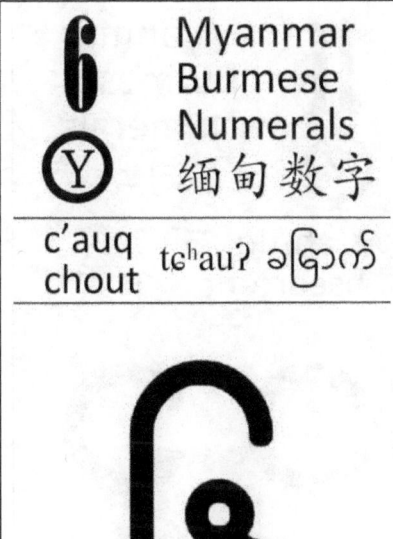

kʰùɴ n̥iʔ

ခုနှစ်

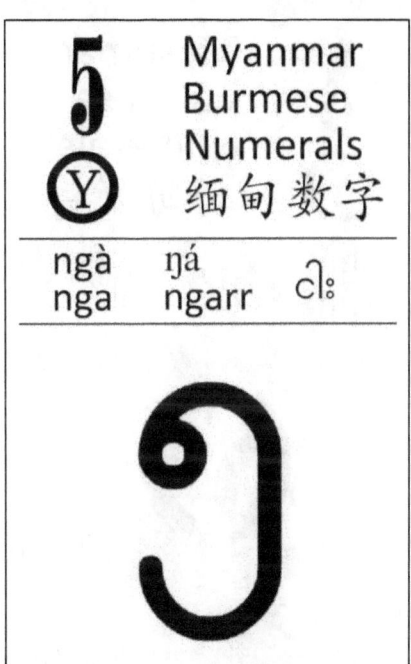

၇

8 Ⓨ
Myanmar
Burmese
Numerals
缅甸数字

shiq
shit
ʃiʔ
ရှစ်

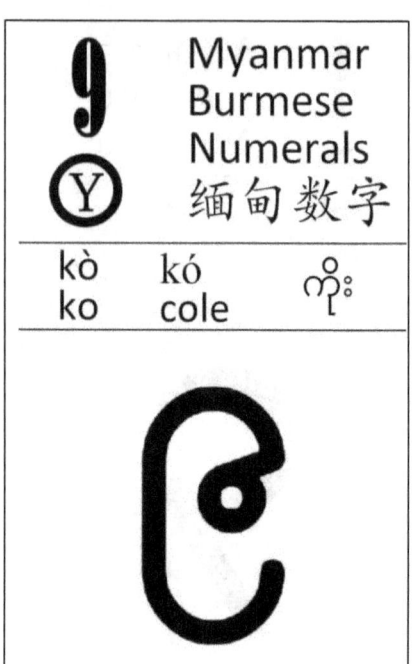

၈

9 Ⓨ
Myanmar
Burmese
Numerals
缅甸数字

kò
ko
kó
cole
ကိုး

၉

10 Ⓨ
Myanmar
Burmese
Numerals
缅甸数字

ta-s'eh
ta-hseare
$s^h\grave{\varepsilon}$
tasal
ဆယ်

၁၀

11 Ⓨ
Myanmar
Burmese
Numerals
缅甸数字

s'eh-tiq
hseare-tit

၁၁

12 Ⓨ

Myanmar
Burmese
Numerals
缅甸数字

s’eh-hniq
hseare-hnit

၁၂

13 Ⓨ

Myanmar
Burmese
Numerals
缅甸数字

s’eh-thòun
hseare-thone

၁၃

14 Ⓨ

Myanmar
Burmese
Numerals
缅甸数字

s’eh-lè
hseare-lay

၁၄

15 Ⓨ

Myanmar
Burmese
Numerals
缅甸数字

s’eh-ngà
hseare-nga

၁၅

16 Ⓨ

Myanmar
Burmese
Numerals
缅甸数字

s'eh-c'auq
hseare-chout

၁၆

17 Ⓨ

Myanmar
Burmese
Numerals
缅甸数字

s'eh-k'u-hniq
hseare khoon-nit

၁၇

18 Ⓨ

Myanmar
Burmese
Numerals
缅甸数字

s'eh-shiq
hseare-shit

၁၈

19 Ⓨ

Myanmar
Burmese
Numerals
缅甸数字

s'eh-kò
hseare-ko

၁၉

20 Ⓨ Myanmar Burmese Numerals 缅甸数字	30 Ⓨ Myanmar Burmese Numerals 缅甸数字
hnuh-s'eh knit sel hna-hseare	thone sel
၂၀	၃၀

Khmer Numerals 柬埔寨高棉数字

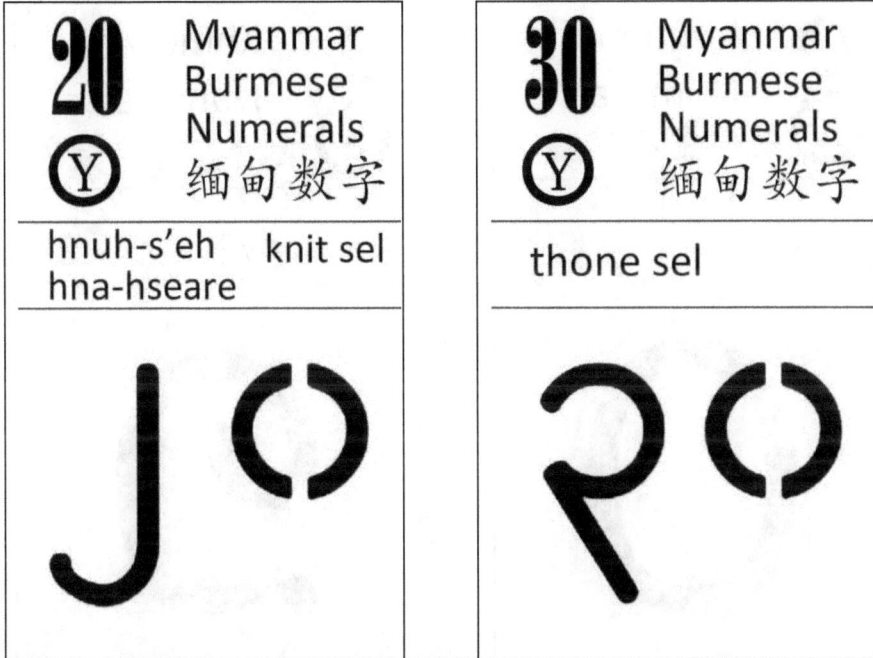

Khmer language is spoken by more than 16 million people in Cambodia as well as by more than 1.3 million in southeastern Thailand and over a million in Vietnam. All of the neighbouring languages, Vietnamese, Thai, Lao and Burmese are tonal languages but Khmer is not. Nevertheless Khmer has been grouped together with those languages as a member of the Austroasiatic language family which stretches from Vietnam to eastern India. The Khmer script, written left to right, developed before the 7th century from the Pallava script of southern India which in turn evolved from the Brahmi script. The earliest Khmer inscriptions are from the 7th century. Khmer numerals are used in a decimal positional notation system derived from India. Cambodia's Sambor inscription of 683 AD, which shows the number 605, provides the earliest epigraphic evidence of the use of zero as a numerical figure . Here zero is represented by a dot as it is in the numerals which Arabic speaking countries borrowed from India.

0 Khmer Numerals
Ⓚ 柬埔寨高棉数字

sony

០

1 Khmer Numerals
Ⓚ 柬埔寨高棉数字

muŏy

១

2 Khmer Numerals
Ⓚ 柬埔寨高棉数字

pir

២

3 Khmer Numerals
Ⓚ 柬埔寨高棉数字

bei

៣

4 Khmer Numerals
Ⓚ 柬埔寨高棉数字

buŏn

៤

5 Khmer Numerals
Ⓚ 柬埔寨高棉数字

prăm

៥

6 Khmer Numerals
Ⓚ 柬埔寨高棉数字

prăm muŏy 5+1

៦

7 Khmer Numerals
Ⓚ 柬埔寨高棉数字

prăm pir 5+2

៧

8 Khmer Numerals
Ⓚ 柬埔寨高棉数字

prăm bey 5+3

៨

9 Khmer Numerals
Ⓚ 柬埔寨高棉数字

prăm buŏn 5+4

៩

10 Khmer Numerals
Ⓚ 柬埔寨高棉数字

dáb

១០

11 Khmer Numerals
Ⓚ 柬埔寨高棉数字

dáb muŏy 10+1
(muŏy dôndáb)

១១

12 Khmer Numerals
Ⓚ 柬埔寨高棉数字

dáb pir 10+2
(pir dôndáb)

១២

13 Khmer Numerals
Ⓚ 柬埔寨高棉数字

dáb bey 10+3
(bei dôndáb)

១៣

14 Khmer Numerals
Ⓚ 柬埔寨高棉数字

dáb buŏn 10+4
(buŏn dôndáb)

១៤

15 Khmer Numerals
Ⓚ 柬埔寨高棉数字

dáb prăm 10+5
(prăm dôndáb)

១៥

16 Khmer Numerals
Ⓚ 柬埔寨高棉数字

dáb prăm muŏy 10+5+1
(prăm muŏy dôndáb)

១៦

17 Khmer Numerals
Ⓚ 柬埔寨高棉数字

dáb prăm pir 10+5+2
(prăm pir dôndáb)

១៧

18 Khmer Numerals
Ⓚ 柬埔寨高棉数字

dáb prăm bey 10+5+3
(prăm bei dôndáb)

១៨

19 Khmer Numerals
Ⓚ 柬埔寨高棉数字

dáb prăm buŏn 10+5+4
(prăm buŏn dôndáb)

១៩

20 Khmer Numerals
Ⓚ 柬埔寨高棉数字

mphey (contraction of muŏy 1 + phey 20)

២០

Thai Numerals 泰国数字

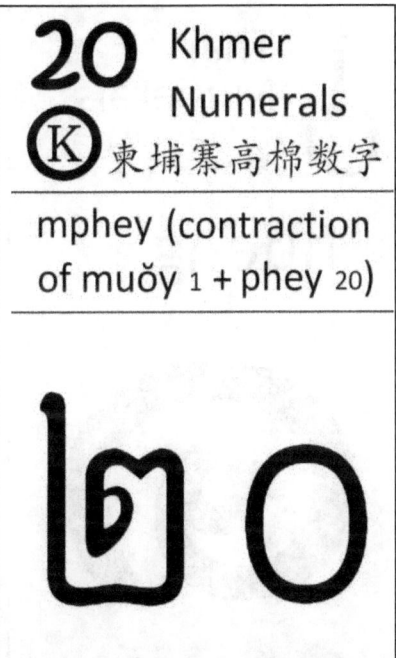

Thai is a tonal language with over 65 million speakers and is mutually intelligible with Lao which has a slightly different script. Thai uses a modified version of the Khmer script which was adapted from the Pallava and Brahmi scripts of India. Thai script is syllabic and can also be used to write Pali and Sanskrit. Over half the words in Thai were borrowed from Sanskrit, Pali and Old Khmer. Inscriptions in Thai first appeared in 1292. Thai numerals are used in a decimal positional notation system derived from India but nowadays the Western Indo-Arabic numerals are also used in Thailand in common with the rest of the world.

0 Thai Numerals
泰国数字

sun

๐

1 Thai Numerals
泰国数字

nueng

๑

2 Thai Numerals
泰国数字

song

๒

3 Thai Numerals
泰国数字

sam

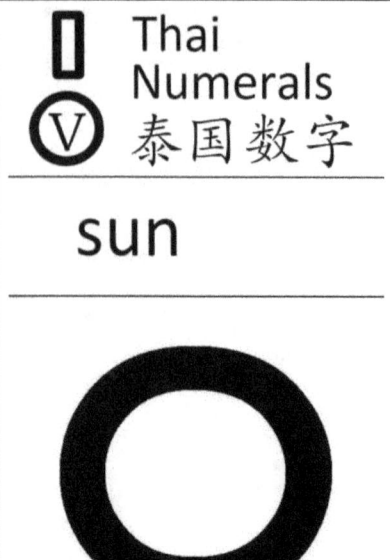

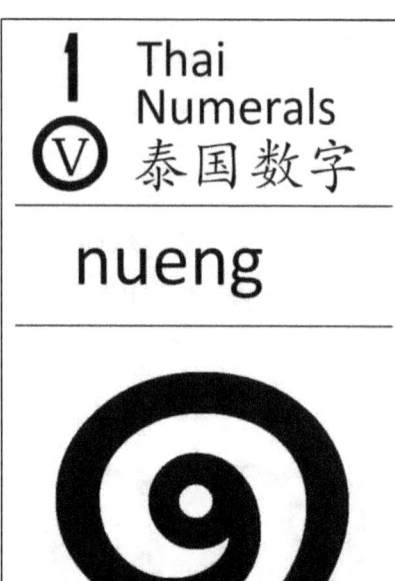

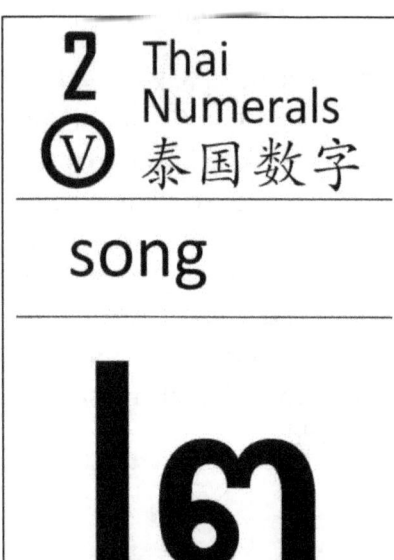

๓

4 Thai Numerals
泰国数字

si

๔

5 Thai Numerals
泰国数字

ha

๕

6 Thai Numerals
泰国数字

hok

๖

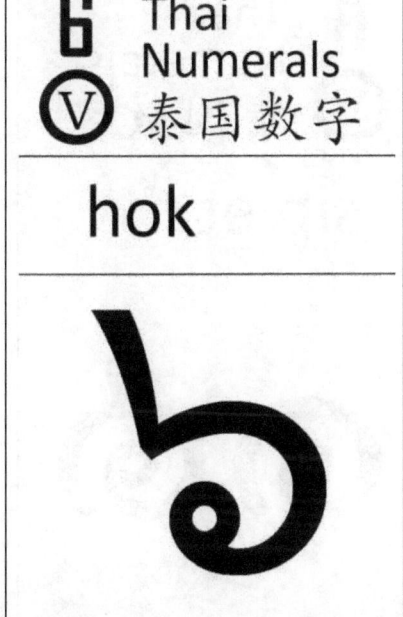

7 Thai Numerals
泰国数字

chet

๗

8 Thai Numerals
Ⓥ 泰国数字

paet

๘

9 Thai Numerals
Ⓥ 泰国数字

kao

๙

10 Thai Numerals
Ⓥ 泰国数字

sip

๑๐

11 Thai Numerals
Ⓥ 泰国数字

sip et

๑๑

12 Thai Numerals
Ⓥ 泰国数字

sip song

๑๒

13 Thai Numerals
Ⓥ 泰国数字

sip sam

๑๓

14 Thai Numerals
Ⓥ 泰国数字

sip si

๑๔

15 Thai Numerals
Ⓥ 泰国数字

sip ha

๑๕

16 Ⓥ Thai Numerals 泰国数字

sip hok

๑๖

17 Ⓥ Thai Numerals 泰国数字

sip chet

๑๗

18 Ⓥ Thai Numerals 泰国数字

sip paet

๑๘

19 Ⓥ Thai Numerals 泰国数字

sip kao

๑๙

20 Thai Numerals Ⓥ 泰国数字	30 Thai Numerals Ⓥ 泰国数字
yi sip	sam sip
๒๐	๓๐

Lao Numerals 老挝数字

The Lao script was derived from the Khmer script and adapted to the Lao language. The Khmer script in turn evolved from the Pallava script of southern India, a variant of the Grantha script which descended from the Brahmi script. The Lao and Thai scripts and languages are mutually comprehensible to educated people in both countries but the Lao alphabet has fewer letters and they are written in a more curvilinear style. Like Thai and Khmer, Lao is written left to right. The Lao numerals one to nine and zero form a decimal positional notation system similar to Khmer and Thai numerals and most modern Indian numerals.

0 Ⓛ	Lao Numerals 老挝数字
sun ສູນ	
໐	

1 Ⓛ	Lao Numerals 老挝数字
nueng ນຶ່ງ	
໑	

2 Ⓛ	Lao Numerals 老挝数字
song ສອງ	
໒	

3 Ⓛ	Lao Numerals 老挝数字
sam ສາມ	
໓	

4 Ⓛ Lao Numerals 老挝数字

si ສີ່

໔

5 Ⓛ Lao Numerals 老挝数字

ha ຫ້າ

໕

6 Ⓛ Lao Numerals 老挝数字

hok ຫົກ

໖

7 Ⓛ Lao Numerals 老挝数字

chet ເຈັດ

໗

8 Ⓛ Lao Numerals 老挝数字

paet ແປດ

໘

9 Ⓛ Lao Numerals 老挝数字

kaw ເກົ້າ

໙

10 Ⓛ Lao Numerals 老挝数字

sip ສິບ

໑໐

11 Ⓛ Lao Numerals 老挝数字

ສິບເອັດ

໑໑

12 Ⓛ
Lao
Numerals
老挝数字

sipsong ສິບສອງ

໑໒

13 Ⓛ
Lao
Numerals
老挝数字

sipsam ສິບສາມ

໑໓

14 Ⓛ
Lao
Numerals
老挝数字

sipsi ສິບສີ່

໑໔

15 Ⓛ
Lao
Numerals
老挝数字

sipha ສິບຫ້າ

໑໕

16 Ⓛ Lao Numerals 老挝数字

siphok ສິບຫົກ

໑໖

17 Ⓛ Lao Numerals 老挝数字

sipchet ສິບເຈັດ

໑໗

18 Ⓛ Lao Numerals 老挝数字

sippaet ສິບແປດ

໑໘

19 Ⓛ Lao Numerals 老挝数字

sipkao ສິບເກົ້າ

໑໙

20 Ⓛ Lao Numerals 老挝数字	30 Ⓛ Lao Numerals 老挝数字
Xao ຊາວ	samsip
໒໐	໓໐

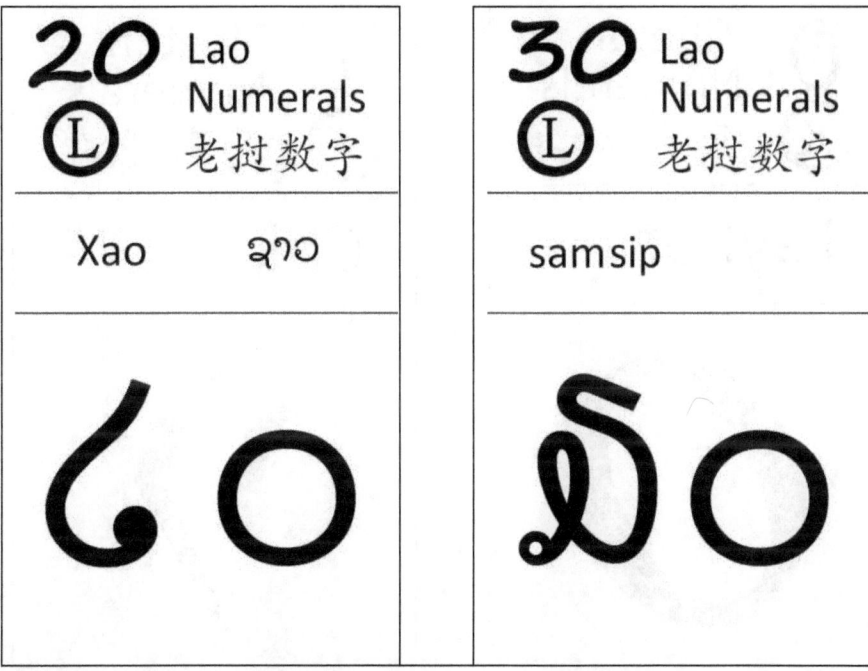

Tibetan Numerals 西藏数字

Tibetan is the official language of the Tibet Autonomous Region of the People's Republic of China and is based on the language of Lhasa, the regional capital. There are nearly 8 million speakers of Tibetan, most living in Tibet but small minorities in Nepal, Sikkim, Bhutan and India. Tibetan is a tonal language and is a member of the Sino-Tibetan language family. The Tibetan alphabet is of Indic origin and was introduced in the mid-7th century. The Limbu and Lepcha alphabets of Sikkim are derived from the Tibetan alphabet. Tibetan has numerals one to nine and zero which form a decimal positional notation system.

0 Ⓧ

Tibetan Numerals
西藏数字

Laykor

༠

1 Ⓧ

Tibetan Numerals
西藏数字

Chig

༡

2 Ⓧ

Tibetan Numerals
西藏数字

Nyi

༢

3 Ⓧ

Tibetan Numerals
西藏数字

Sum

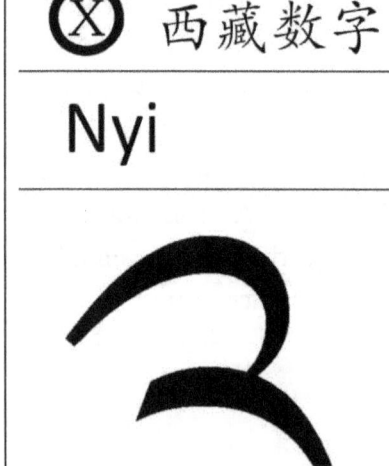

༣

4 ⓧ Tibetan Numerals 西藏数字

Shi

༤

5 ⓧ Tibetan Numerals 西藏数字

Nga

༥

6 ⓧ Tibetan Numerals 西藏数字

Trug

༦

7 ⓧ Tibetan Numerals 西藏数字

Dün

༧

8 Tibetan Numerals
ⓧ 西藏数字

Gyay

༨

9 Tibetan Numerals
ⓧ 西藏数字

Gu

༩

10 Tibetan Numerals
ⓧ 西藏数字

Chu

༡༠

11 Tibetan Numerals
ⓧ 西藏数字

Chu-chig

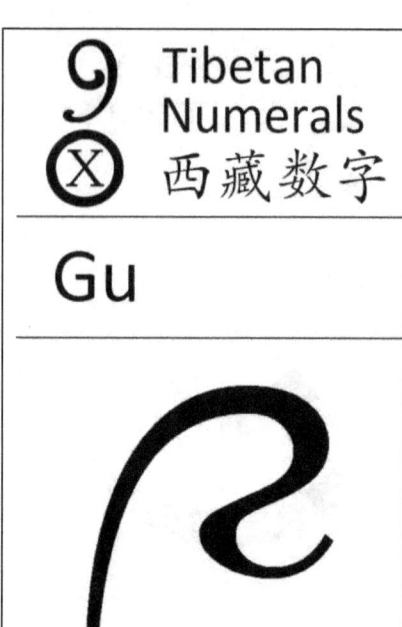

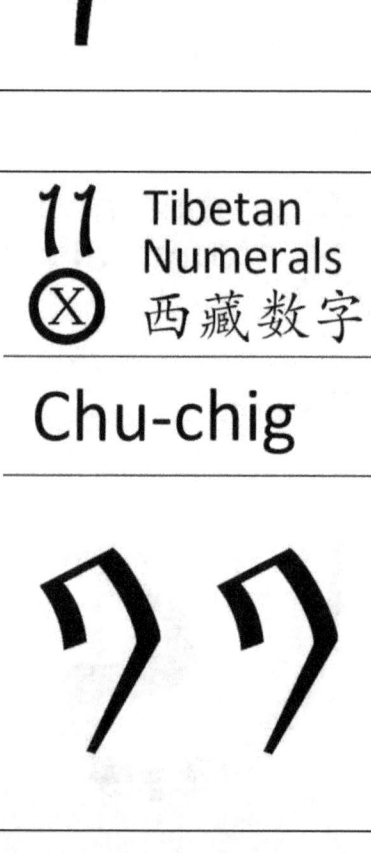

༡༡

12 Tibetan Numerals
Ⓧ 西藏数字

Chu-nyi

༡༢

13 Tibetan Numerals
Ⓧ 西藏数字

Chu-sum

༡༣

14 Tibetan Numerals
Ⓧ 西藏数字

Chu-shi

༡༤

15 Tibetan Numerals
Ⓧ 西藏数字

Chu-nga

༡༥

16 Tibetan Numerals
Ⓧ 西藏数字

Chu-trug

༡༦

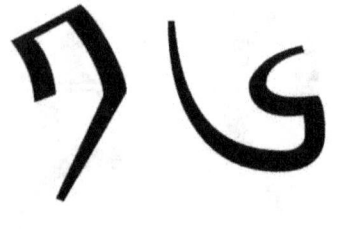

17 Tibetan Numerals
Ⓧ 西藏数字

Chu-dün

༡༧

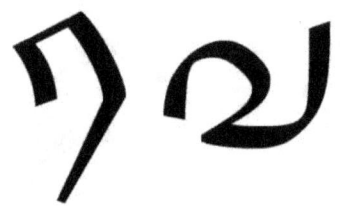

18 Tibetan Numerals
Ⓧ 西藏数字

Chu-gyay

༡༨

19 Tibetan Numerals
Ⓧ 西藏数字

Chu-gu

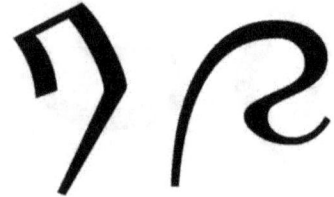

༡༩

20 Tibetan Numerals
ⓧ 西藏数字

Nyi-shu

༢༠

Mongolian Numerals 蒙古数字

Mongolian is a member of the Altaic language family and is spoken by more than 5 million people in Mongolia, Inner Mongolia, Russia and Afghanistan. The Khalkha dialect is the national language of Mongolia and the Oirat, Chahar and Ordos dialects are spoken in Inner Mongolia. The Buryat and Kalmyk languages spoken in Russia and the Moghul language in Afghanistan are closely related. Around 1204 Genghis Khan captured the Uyghur scribe Tata Tonga who created the Mongolian alphabet based on the Old Uyghur alphabet which was of Syriac origin. Mongolian script is written in vertical columns from top to bottom and left to right and still used in China'a Inner Mongolia. In February 1941 Mongolia abolished the traditional alphabet and introduced a Latin alphabet which was soon abandoned in favour of the Cyrillic alphabet. Mongolian numerals are related to and based on Tibetan numerals.

0 Mongolian Numerals
Ⓦ 蒙古数字

teg

᠐

1 Mongolian Numerals
Ⓦ 蒙古数字

neg

᠑

2 Mongolian Numerals
Ⓦ 蒙古数字

khoior

᠒

3 Mongolian Numerals
Ⓦ 蒙古数字

gurav

᠓

4 Mongolian Numerals
蒙古数字

döröv

᠔

5 Mongolian Numerals
蒙古数字

tav

᠕

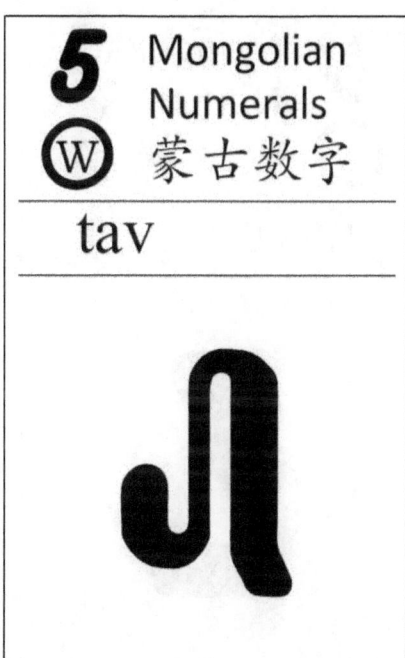

6 Mongolian Numerals
蒙古数字

zurgaa

᠖

7 Mongolian Numerals
蒙古数字

doloo

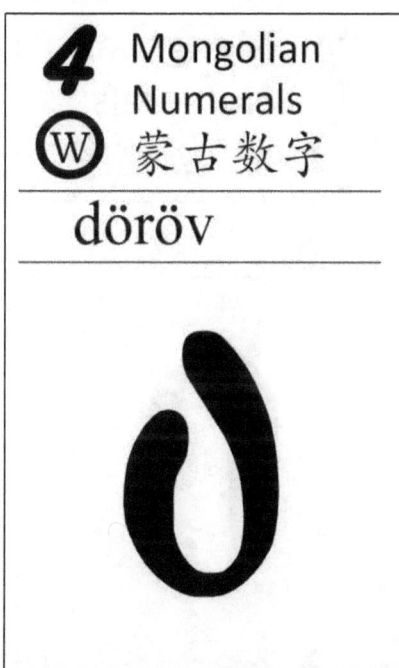

᠗

8 Mongolian Numerals 蒙古数字

naym

᠘

9 Mongolian Numerals 蒙古数字

es

᠙

10 Mongolian Numerals 蒙古数字

arav arvan

᠑᠐

11 Mongolian Numerals 蒙古数字

arvan neg

᠑᠑

12 Mongolian Numerals
蒙古数字

arvan khoior

᠑᠒

13 Mongolian Numerals
蒙古数字

arvan gurav

᠑᠓

14 Mongolian Numerals
蒙古数字

arvan dorov

᠑᠔

15 Mongolian Numerals
蒙古数字

arvan tav

᠑᠕

16 Mongolian Numerals 蒙古数字

arvan zurgaa

᠑᠖

17 Mongolian Numerals 蒙古数字

arvan doloo

᠑᠗

18 Mongolian Numerals 蒙古数字

arvan naim

᠑᠘

19 Mongolian Numerals 蒙古数字

arvan es

᠑᠙

20 Mongolian Numerals

Ⓦ 蒙古数字

horv

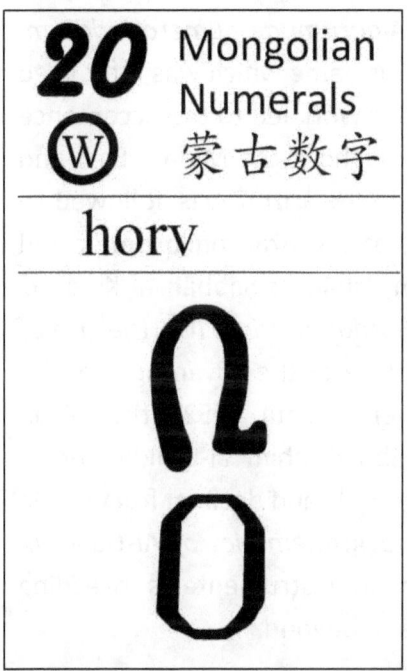

Eastern Arabic Numerals 东阿拉伯数字

Numerals used in Arabic speaking countries were adapted from Indian numerals which developed around the middle of the 3rd century BC and continued to evolve until the 7th century when they were known as Devanagari numerals. In 662 AD the Syrian Nestorian scholar Severus Sebokht wrote about the advantages of the Indian system which employed symbols for the numerals one to nine and zero. In 628 the Indian astronomer and mathematician Brahmagupta (c. 598-after 628) in his *Brāhmasphuṭasiddhānta* wrote about negative and positive numbers and gave rules for the computation of zero. The first indisputable use of zero as a numeral in India can be seen in the Gwalior inscription dated 876 showing the number 270. The fully developed system of numerals had emerged by the 8th or 9th century and is described in about 825 by the Persian scholar Muḥammad ibn Mūsā al-Khwārizmī (c. 780-c. 850) in his book *On the Calculation with Hindu Numerals* which was translated into Latin under the title *Algoritmi de numero Indorum*. Algoritmi was the Latin translation of his name which was later used in the term "algorithm". Al-Khwārizmī's work led to the acceptance of the Indian system of numerals throughout the Middle East and eventually in Europe. The work by al-Khwārizmī was followed in about 830 by the four volume book of the Arab philosopher and mathematician Abu Yūsuf Ya'qūb ibn 'Isḥāq aṣ-Ṣabbāḥ al-Kindī (c. 801- 873) entitled *Ketab fi Isti'mal al-'Adad al-Hindi (On the Use of the Indian Numerals)* which spoke further of the advantages of the new system of numeration. Then about 952 the Arab mathematician Abu'l Hasan Ahmad ibn Ibrahim al-Uqlidisi wrote about the positional use of Arabic numerals and decimal fractions in *Kitab al-Fusul fi al-Hisab al-Hindi (The Arithemetics of Al-Uqlidisi).* Al-Khwārizmī, al-Kindī and al-Uqlidisi were instrumental is spreading the new numerals to the Arab world and beyond.

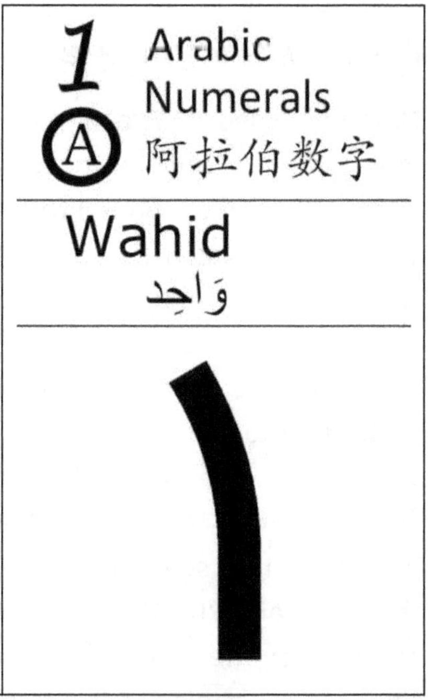

2 Ⓐ Arabic Numerals 阿拉伯数字

Itnan

اثْنَان

٢

3 Ⓐ Arabic Numerals 阿拉伯数字

Talatah

ثَلَاثَة

٣

4 Ⓐ Arabic Numerals 阿拉伯数字

Arab'ah

أَرْبَعَة

٤

5 Ⓐ Arabic Numerals 阿拉伯数字

Hamsah

خَمْسَة

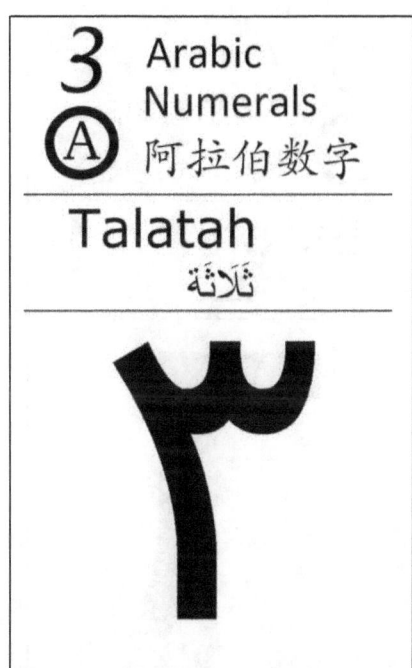

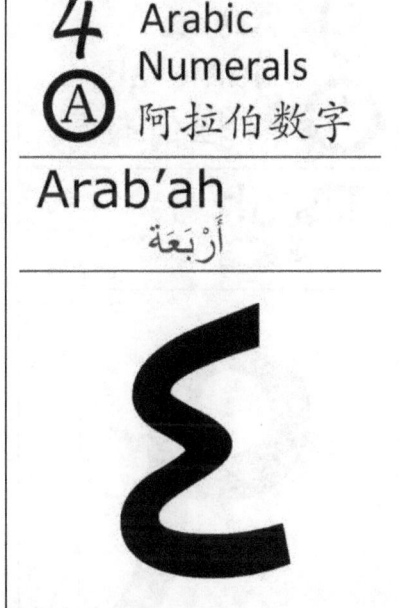

٥

6 Ⓐ Arabic Numerals 阿拉伯数字

Sittah

سِتَّة

٦

7 Ⓐ Arabic Numerals 阿拉伯数字

Sab'ah

سَبْعَة

٧

8 Ⓐ Arabic Numerals 阿拉伯数字

Tamaniyyah

ثَمَانِيَة

٨

9 Ⓐ Arabic Numerals 阿拉伯数字

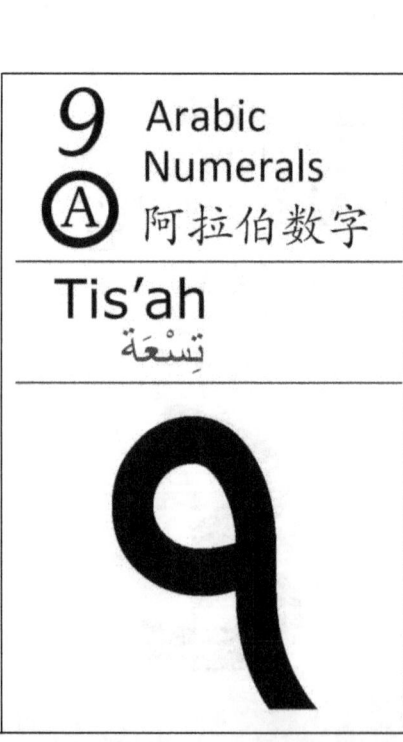

Tis'ah

تِسْعَة

٩

10 Arabic Numerals
Ⓐ 阿拉伯数字

Asarah
عَشْرَة

١٠

11 Arabic Numerals
Ⓐ 阿拉伯数字

Hidashar
عَشَر أَحَد

١١

12 Arabic Numerals
Ⓐ 阿拉伯数字

Itnashar
عَشَر اثْنَا

١٢

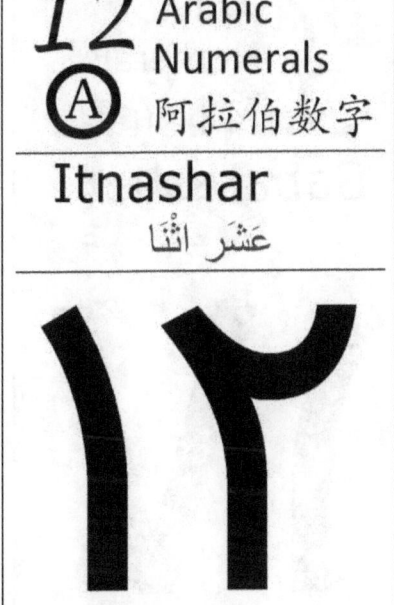

13 Arabic Numerals
Ⓐ 阿拉伯数字

Talatashar
عَشَر ثَلَاثَة

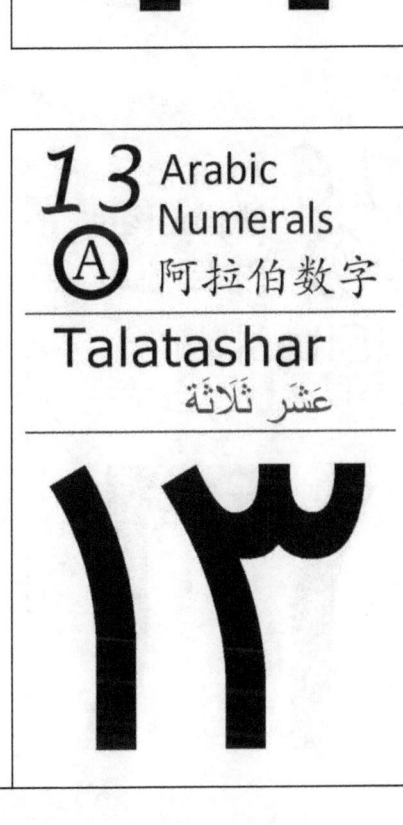

١٣

14 Arabic Numerals
Ⓐ 阿拉伯数字

Arbatashar
عَشْرَة أَرْبَعَة

١٤

15 Arabic Numerals
Ⓐ 阿拉伯数字

Kamastashar
عَشَر خَمْسَة

١٥

16 Arabic Numerals
Ⓐ 阿拉伯数字

Sitashar
عَشْرَة سَت

١٦

17 Arabic Numerals
Ⓐ 阿拉伯数字

Sabatashar
عَشَر سَبْعَة

١٧

18 Arabic Numerals
Ⓐ 阿拉伯数字

Tamantashar
ثَمَانِيَة عَشَر

١٨

19 Arabic Numerals
Ⓐ 阿拉伯数字

Tisatashar
تِسْعَة عَشَر

١٩

20 Arabic Numerals
Ⓐ 阿拉伯数字

Ishrin
عِشْرُوْن

٢٠

30 Arabic Numerals
Ⓐ 阿拉伯数字

Thalathun
ثلاثون

٣٠

Persian Numerals 波斯数字

Persian, also called Farsi, is the official language of Iran and is a member of the Indo-Iranian branch of the Indo-European language family. Two closely related Persian languages, Dari and Tajik, are official languages in Afghanistan and Tajikistan. Old Persian, the language of the Achaemenid Empire, was written in cuneiform script and spoken until the 3rd century BC. Middle Persian, known as Pahlavi, was the language of the Sasanian Empire. It was written in the Aramaic script and spoken from the 3rd century BC until the 9th century. Modern Persian is written in a modified version of the Arabic alphabet. The Persian numerals for 4, 5 and 6 are different from the standard numerals used in Arab countries. Persian numerals are also used in the writing of Urdu.

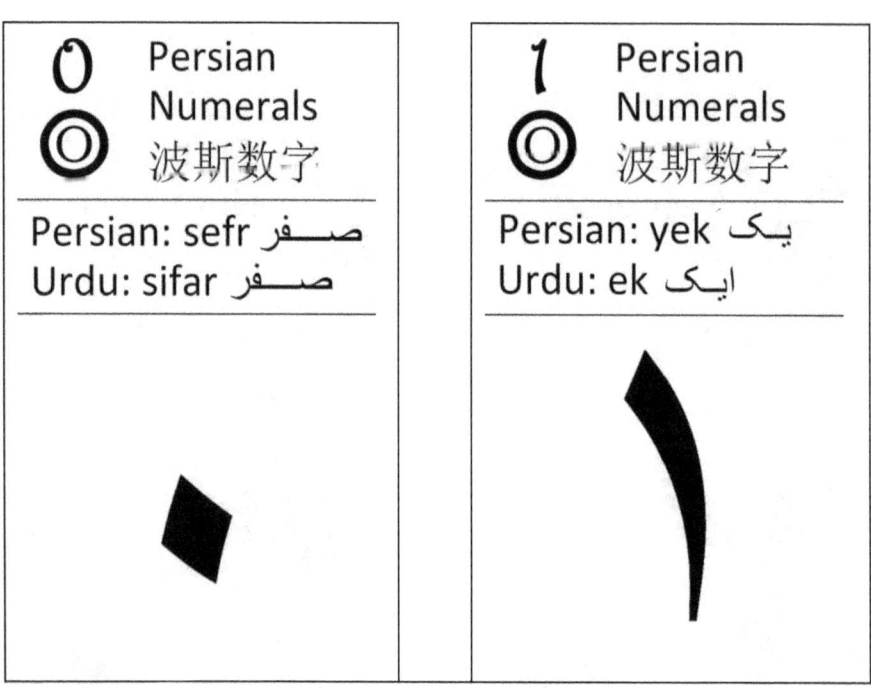

0 Persian Numerals 波斯数字

Persian: sefr صفر
Urdu: sifar صفر

٠

1 Persian Numerals 波斯数字

Persian: yek یک
Urdu: ek ایک

١

2 Persian Numerals 波斯数字

Persian: do دو
Urdu: do دو

۲

3 Persian Numerals 波斯数字

Persian: seh سه
Urdu: tīn تیـــن

۳

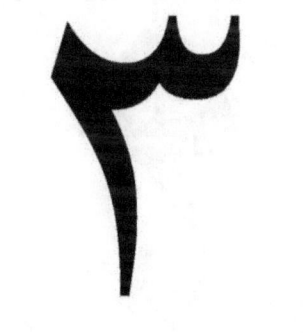

4 Persian Numerals 波斯数字

Persian: chahaar چهار
Urdu: chār چار

۴

5 Persian Numerals 波斯数字

Persian: panj پنـــج
Urdu: pānch پـــانچ

۵

6 ◎ Persian Numerals 波斯数字

Persian: shesh شش
Urdu: chhah سات

۶

7 ◎ Persian Numerals 波斯数字

Persian: haft هفت
Urdu: sāt سات

۷

8 ◎ Persian Numerals 波斯数字

Persian: hasht هشت
Urdu: āth آٹھ

۸

9 ◎ Persian Numerals 波斯数字

Persian: noh نه
Urdu: nau نو

۹

10 ◎ Persian Numerals 波斯数字

Persian: dah ده
Urdu: das دس

۱۰

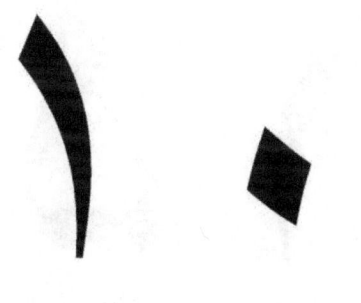

11 ◎ Persian Numerals 波斯数字

Persian: yazdah
Urdu: gyārah

۱۱

12 ◎ Persian Numerals 波斯数字

Persian: davaazdah
Urdu: bārah

۱۲

13 ◎ Persian Numerals 波斯数字

Persian: sizdah
Urdu: tērah

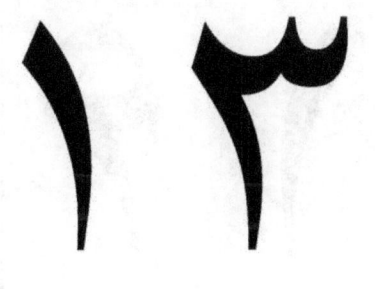

۱۳

14 Persian Numerals 波斯数字

Persian: chahaardah
Urdu: chaudah

۱۴

15 Persian Numerals 波斯数字

Persian: poonzdah
Urdu: pamdrah

۱۵

16 Persian Numerals 波斯数字

Persian: shoonzdah
Urdu: solah

۱۶

17 Persian Numerals 波斯数字

Persian: hivdah
Urdu: satrah

۱۷

18 Persian Numerals 波斯数字

Persian: hijdah
Urdu: atthārah

۱۸

19 Persian Numerals 波斯数字

Persian: noonzdah
Urdu: unnis

۱۹

20 Persian Numerals 波斯数字

Persian: bist
Urdu: bīs

۲۰

30 Persian Numerals 波斯数字

Persian: si
Urdu: tīs

۳۰

The Introduction of Western Arabic Numerals to Europe

The system of numerals developed in India around 500 AD used 10 symbols including zero. Arab mathematicians in Bagdad adopted this system and it was transmitted to Arabic speaking regions in the western Mediterranean including Libya, Algeria, Morocco and Arab Spain, known then as Al Andalus. New variants of the numerals had developed in these countries by the 10th century and they should more correctly be called Western Arabic Numerals to distinguish them from the Eastern Arabic Numerals developed in Bagdad where they are called Indian numbers in acknowledgement of their Indian origin.

These Western variants, first mentioned in the *Codex Vigilanus* of 976, were introduced to Spain around 900. In the 980s Gilbert of Aurillac, who had studied in Spain as a young man and later became Pope Sylvester II, passed on his knowledge of the numerals to scholars in Europe. Then Leonardo Fibonacci, a mathematician from the Republic of Pisa, in his book *Liber Abaci* of 1202, publicized the Western Arabic numerals which he had studied while in Bejaia, Algeria. However, it was not until the 15th century that the new numerals began to gain acceptance in Europe and only by the mid 16th century were they in common use. The printing press in Europe helped popularize the new numerals and standardize the symbols used to write them. In Western countries the term Western Indo-Arabic Numerals is too long and cumbersome leaving us with Arabic Notation as the only reasonable though ambiguous name for our numbers.

I have selected five examples of European Medieval numerals from the 10th to the 13th century appearing in a table created by the French scholar Jean-Étienne Montucla in his book of 1757 entitled *Histoire de la Mathematique*. This table appears in *Arabic Numerals* (https://en.wikipedia.org/wiki/Arabic_numeral) published on the internet by Wikipedia. As the numerals in this table are small and of relatively low DPI, I decided to take several days to redraw them in a much larger size keeping as near as possible to the original pattern and style.

However, there will inevitably be differences and variations from the originals but the larger size will hopefully allow readers to appreciate the beginnings of a new system of numeration. My experience in studying and teaching Asian languages has convinced me that letters, characters and numerals in foreign languages should be presented in a much larger size than native speakers would use.

The first nine numbers below are from the *Codex Vigilanus* of 976 which Montucla accessed in the Library of San Lorenzo del Escorial in Spain. 1, 6, 7, 8 and 9 look much like our modern Western Arabic numerals although 6 also resembles a lower case B. The *Vigilanus* 2 is an upside-down version of our modern two. Number 3 has three almost horizontal lines much like in the Indian Brahmi numerals but joined together. The *Vigilanus* 5 resembles the Sanskrit Devanagari 5 and the 5 in Gujarati, Gurmukhi and Tibetan while the *Vigilanus* 4 somewhat resembles the 4 in Malayalam. This demonstrates further the Indian origin of these numbers.

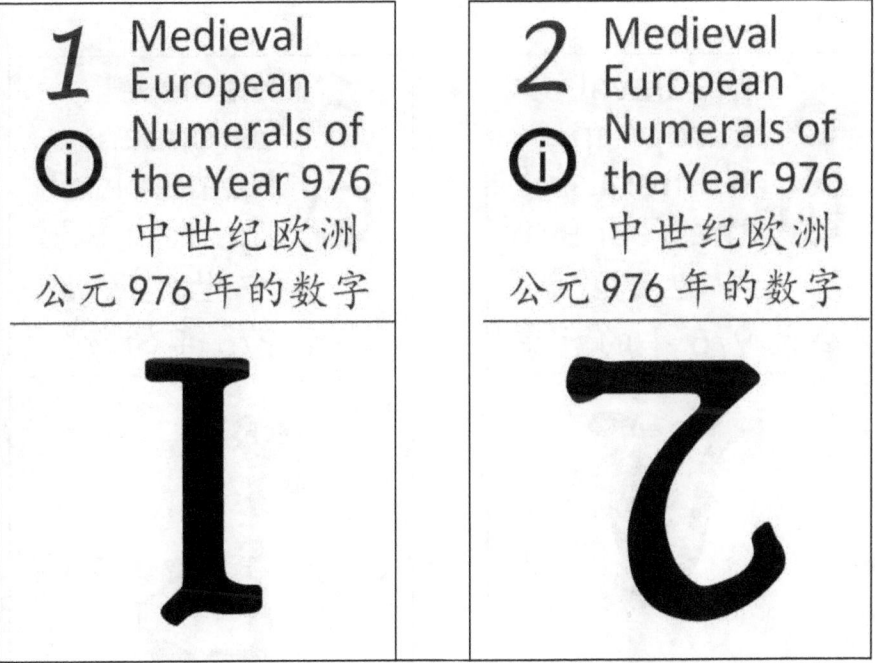

1 Medieval European Numerals of the Year 976 中世纪欧洲公元976年的数字	2 Medieval European Numerals of the Year 976 中世纪欧洲公元976年的数字
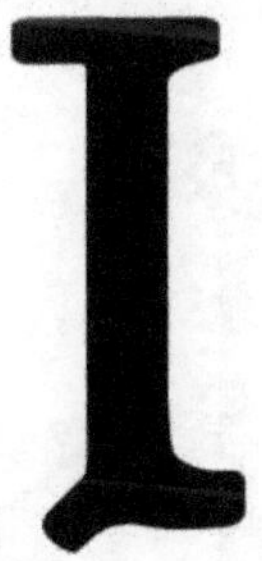	

3 Medieval European Numerals of the Year 976
中世纪欧洲公元 976 年的数字

4 Medieval European Numerals of the Year 976
中世纪欧洲公元 976 年的数字

5 Medieval European Numerals of the Year 976
中世纪欧洲公元 976 年的数字

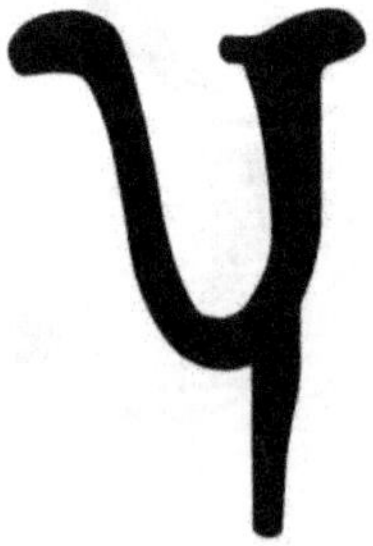

6 Medieval European Numerals of the Year 976
中世纪欧洲公元 976 年的数字

7 ⓘ Medieval European Numerals of the Year 976
中世纪欧洲
公元 976 年的数字

8 ⓘ Medieval European Numerals of the Year 976
中世纪欧洲
公元 976 年的数字

9 ⓘ Medieval European Numerals of the Year 976
中世纪欧洲
公元 976 年的数字

Numerials from Montpellier, France

The following 10 numerals are from the 11th century and were obtained by Jean-Étienne Montucla from the Library of the School of Medicine of Montpellier (Bibliotheque de l'École de Medecine de Montpellier). Here 8 and 9 resemble modern European numerals. 1 is a J and this carries on a tradition in Roman Notation where in some circumstances J is substituted for I. Zero in the form of a circle is used here but it surrounds the letter A. Perhaps this goes back to the use of letters of the alphabet as numerals by the Romans and Greeks. Number 3 shows the three horizontal strokes of Brahmi joined together and 5 and 6 are almost mirror images of one another although 6 looks a little like the letter P. Number 5 also resembles the 5 in Sanskrit Devanagari and in Gujarati, Gurmukhi and Tibetan. Number 7 looks a little like the Eastern Arabic 8 and number 4 appears to be a new creation with little or no similarity to Indian numerals.

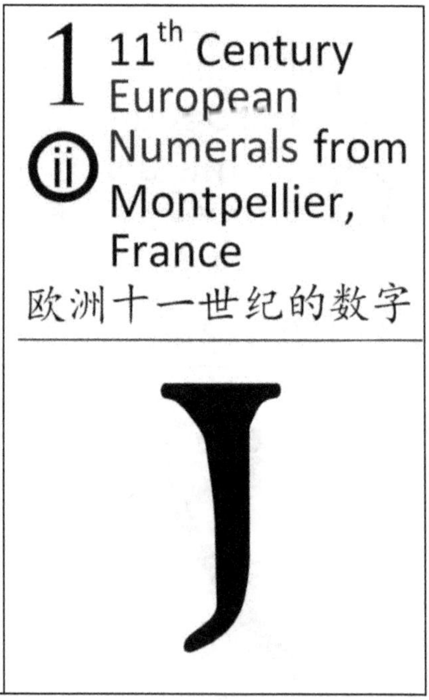

2 11th Century European Numerals from Montpellier, France

(ii)

欧洲十一世纪的数字

3 11th Century European Numerals from Montpellier, France

(ii)

欧洲十一世纪的数字

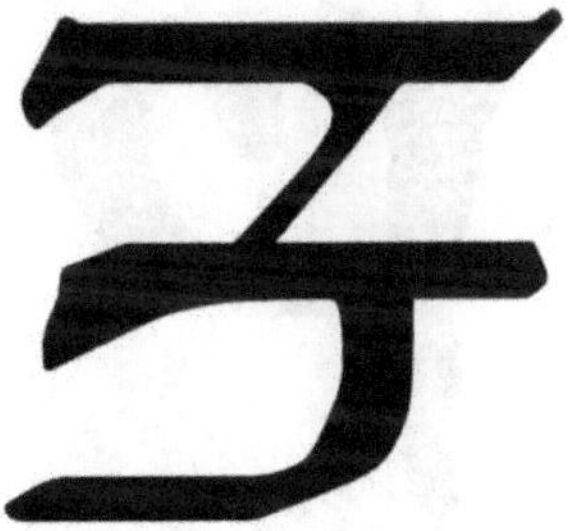

4 11th Century European Numerals from Montpellier, France

(ii)

欧洲十一世纪的数字

5 11th Century European Numerals from Montpellier, France

(ii)

欧洲十一世纪的数字

6 11th Century European Numerals from Montpellier, France

欧洲十一世纪的数字

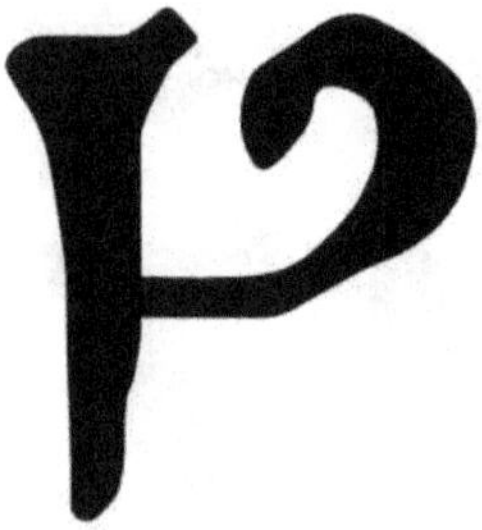

7 11th Century European Numerals from Montpellier, France

欧洲十一世纪的数字

8 11th Century European Numerals from Montpellier, France

欧洲十一世纪的数字

9 11th Century European Numerals from Montpellier, France

欧洲十一世纪的数字

The Late 11th Century European Numerals of Bernelinus

Bernelinus was the French pupil of Gerbert d'Aurillac (c. 946-1003) who later became the first French pope with the reign title Sylvester II. Gerbert had been taken to Barcelona in 967 and later studied in Cordoba and Seville where he acquired a good knowledge of Arab mathematics and the Western Arabic numerals which were still evolving. Later when he became pope he promoted the study of mathematics and astronomy and taught the new numbering system to Bernelinus who must have been very young at the time. In fact the numerals written by Bernelinus may be much earlier than Jean-Étienne Montucla's stated date at the end of the 11th century. The numerals of Bernelinus are not very different from those obtained from the Medical School at Montpellier except that number 9 is lying on its side as if turned clockwise. This is not uncommon in Asia where the 3 in Malayalam, Oriya, Khmer and Thai appear to have been turned anti-clockwise to face downwards.

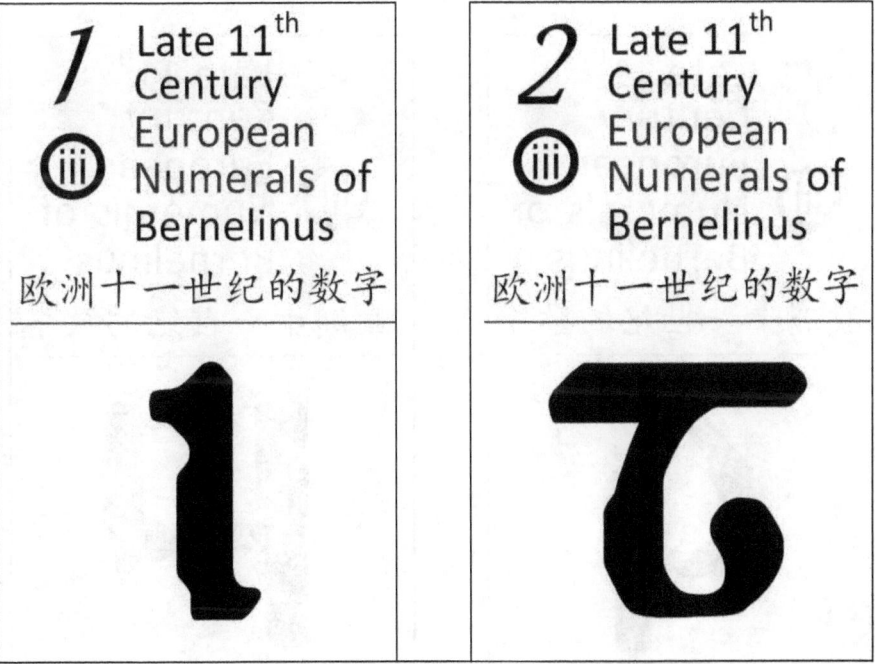

1 Late 11th Century European Numerals of Bernelinus

ⓘ (iii)

欧洲十一世纪的数字

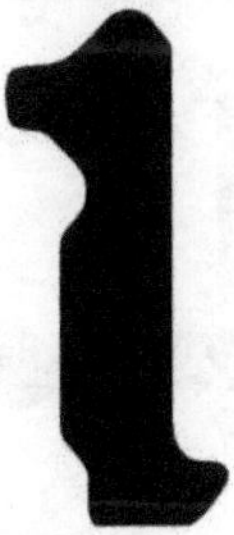

2 Late 11th Century European Numerals of Bernelinus

(iii)

欧洲十一世纪的数字

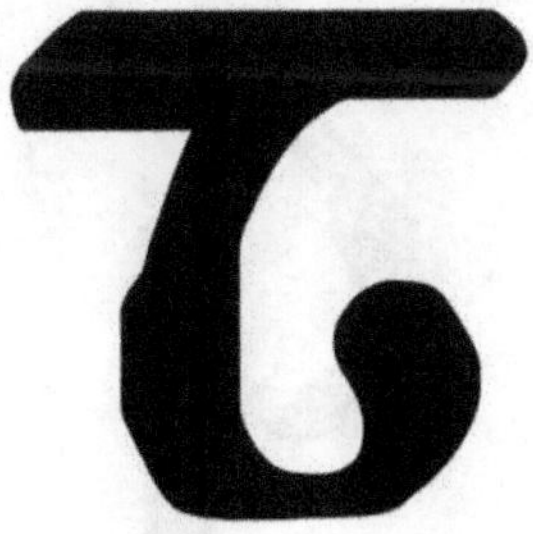

3 Late 11th Century
(iii) European Numerals of Bernelinus

欧洲十一世纪的数字

4 Late 11th Century
(iii) European Numerals of Bernelinus

欧洲十一世纪的数字

5 Late 11th Century
(iii) European Numerals of Bernelinus

欧洲十一世纪的数字

6 Late 11th Century
(iii) European Numerals of Bernelinus

欧洲十一世纪的数字

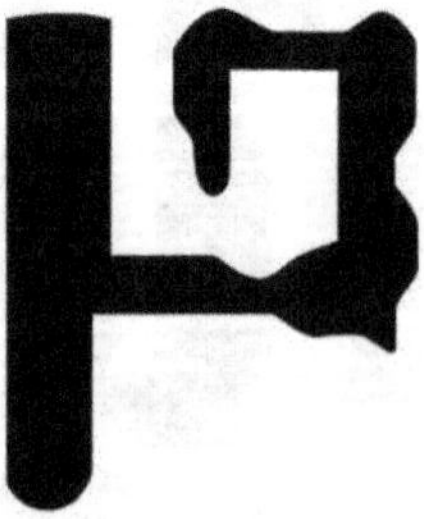

7 Late 11th Century European Numerals of Bernelinus

欧洲十一世纪的数字

8 Late 11th Century European Numerals of Bernelinus

欧洲十一世纪的数字

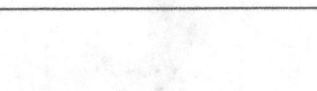

9 Late 11th Century European Numerals of Bernelinus

欧洲十一世纪的数字

The 12th Century European Numerals of Gerlandus

Gerlandus may refer to Garlandus Compotista (Garland the Computist) an eleventh-century logician from the school of Liege. It may also refer to the early 12th century scholar Gerlandus of Besançon. The *Dialectica* published by L. M. de Rijk was probably authored by the latter. Here we see number 3 turned anticlockwise and number 4 looking like an upper case B with a tail. The Gerlandus 7 is much the same as the Eastern Arabic 7 but the 9 has been turned upside-down and looks too much like 6.

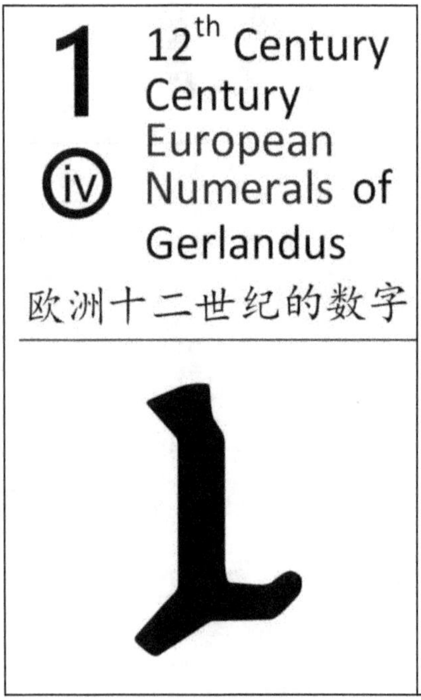

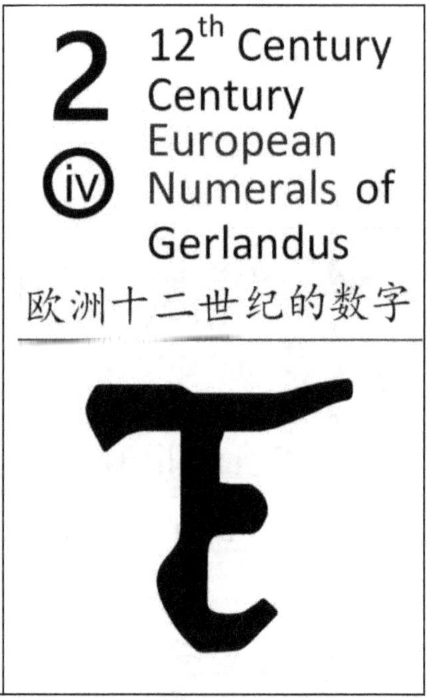

3 (iv) 12th Century Century European Numerals of Gerlandus

欧洲十二世纪的数字

4 (iv) 12th Century Century European Numerals of Gerlandus

欧洲十二世纪的数字

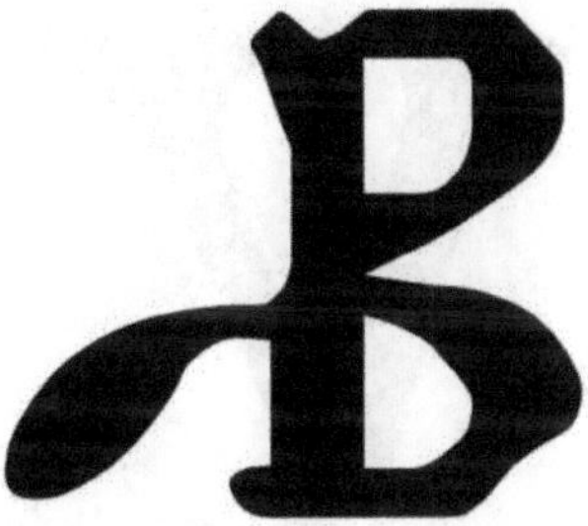

5 (iv) 12th Century Century European Numerals of Gerlandus

欧洲十二世纪的数字

6 (iv) 12th Century Century European Numerals of Gerlandus

欧洲十二世纪的数字

7 12th Century Century European Numerals of Gerlandus

(iv)

欧洲十二世纪的数字

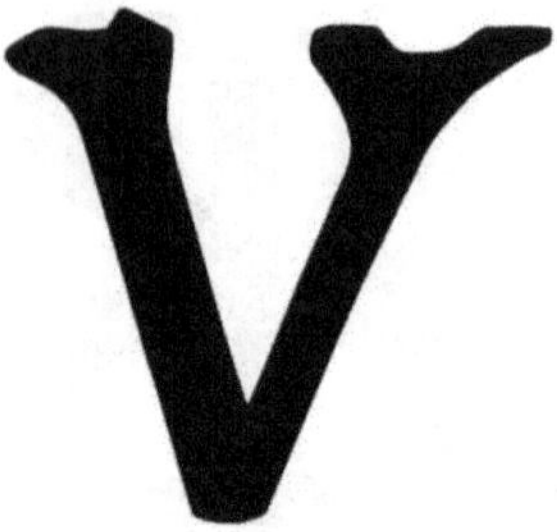

8 12th Century Century European Numerals of Gerlandus

(iv)

欧洲十二世纪的数字

9 12th Century Century European Numerals of Gerlandus

(iv)

欧洲十二世纪的数字

The European Numerals of English Scholar Roger Bacon

These numerals were written by the English philosopher, scientist and Franciscan friar Roger Bacon (1220-1292). Bacon's numbers 1, 3, 6, 8, 9 and 10 are similar to modern European numbers. His number 7 is the same as the Eastern Arabic 8. His number 4 is an inverted version of the Sanskrit Devanagari, Gujarati and Odiya 4.

1 European Numerals of Philosopher Roger Bacon (1220-1292)
ⓥ
欧洲十三世纪的数字

2 European Numerals of Philosopher Roger Bacon (1220-1292)
ⓥ
欧洲十三世纪的数字

3 Ⓥ European Numerals of Philosopher Roger Bacon (1220-1292)

欧洲十三世纪的数字

4 Ⓥ European Numerals of Philosopher Roger Bacon (1220-1292)

欧洲十三世纪的数字

5 Ⓥ European Numerals of Philosopher Roger Bacon (1220-1292)

欧洲十三世纪的数字

6 Ⓥ European Numerals of Philosopher Roger Bacon (1220-1292)

欧洲十三世纪的数字

7 Ⓥ European Numerals of Philosopher Roger Bacon (1220-1292)

欧洲十三世纪的数字

8 Ⓥ European Numerals of Philosopher Roger Bacon (1220-1292)

欧洲十三世纪的数字

9 Ⓥ European Numerals of Philosopher Roger Bacon (1220-1292)

欧洲十三世纪的数字

10 Ⓥ European Numerals of Philosopher Roger Bacon (1220-1292)

欧洲十三世纪的数字

References

Maya and Quechua Numerals

https://en.wikipedia.org/wiki/Quipu; last edited on 10 Sep 2017, at 08:22.

http://www.ancientscripts.com/quipu.html.

https://www.thoughtco.com/introduction-to-quipu-inca-writing-system-172285; *Quipu - South America's Ancient Undeciphered Writing System: An Introduction to the Inca Writing System Known as Quipu* by K. Kris Hirst; Updated April 14, 2017.

https://en.wikipedia.org/wiki/Cangjie; last edited on 30 May 2017, at 18:45; accessed 17 Sep 2017.

http://www.ancientpages.com/2017/03/15/ancient-chinese-version-of-quipu-tradition-of-tying-knots-dates-back-to-antiquity/ "Ancient Chinese Version Of Quipu -Tradition Of Tying Knots Dates Back To Antiquity", *Ancient Pages: Civilizations, Featured Stories, News, March 15, 2017,* written by A. Sutherland, AncientPages.com Staff Writer; accessed 17 Sep 2017.

https://www.britannica.com/technology/quipu.

https://en.wikipedia.org/wiki/Maya_numerals; last edited on 20 August 2017, at 18:40.

http://mathcentral.uregina.ca/RR/database/RR.09.00/hubbard1/MayanNumerals.html. *Mayan Numerals* by Jamie Hubbard.

https://www.easycalculation.com/funny/numerals/mayan.php

https://www.isnare.com/encyclopedia/Maya_numerals. *Free Encyclopedia*. "Wiktionary Definitions"; Accessed 17 Sep 2017.

http://www.look4ward.co.uk/x-files/koh-ker-temple-mayan-pyramid-in-cambodia/. "Koh Ker Temple: Mayan Pyramid In Cambodia", by Lindsey Rees, Look4ward: *Archeology • Architecture • X-Files*, May 8, 2017; accessed 17 Sep 2017.

Chinese Numerals

https://en.wikipedia.org/wiki/Chinese_numerals; last edited on 30 August 2017, at 14:32.

http://www.omniglot.com/chinese/numerals.htm

http://www.omniglot.com/chinese/jiaguwen.htm

http://chinaknowledge.de/Literature/Historiography/oracle.html; July 10, 2010, Ulrich Theobald, *An Encyclopaedia on Chinese History, Literature and Art.*

https://www.thoughtco.com/oracle-bones-shang-dynasty-china-172015; by K. Kris Hirst; Updated February 13, 2017

https://arxiv.org/ftp/arxiv/papers/1511/1511.08033.pdf; On Mathematical Symbols in China, By Fang Li, Yong Zhang
(Department of mathematics, Zhejiang University, Hangzhou).

https://en.wikipedia.org/wiki/Cangjie; last edited on 30 May 2017, at 18:45; accessed 17 Sep 2017.

https://en.wikipedia.org/wiki/Chinese_characters; last edited on 13 September 2017, at 06:36

https://www.brown.edu/about/administration/international-affairs/year-of-china/language-and-cultural-resources/introduction-chinese-characters/introduction-chinese-characters; Text by Yang Wang

https://en.wikipedia.org/wiki/Chinese_numerals; last edited on 30 August 2017, at 14:32.

http://www.mandarintools.com/numbers.html; Chinese Numbers.

https://en.wikipedia.org/wiki/Hindu%E2%80%93Arabic_numeral_system; last edited on 31 August 2017, at 19:46.

Babylonian Numbers

https://en.wikipedia.org/wiki/Babylonian_numerals; last edited on 6 June 2017, at 22:09.

https://www.easycalculation.com/funny/numerals/babylonian.php

https://explorable.com/babylonian-mathematics Martyn Shuttleworth (Jul 19, 2010). Babylonian Mathematics And Babylonian Numerals. Retrieved Sep 22, 2017 from Explorable.com: https://explorable.com/babylonian-mathematics

http://www-groups.dcs.st-and.ac.uk/~history/HistTopics/Babylonian numerals.html; Article by: J J O'Connor and E F Robertson ; JOC/EFR December 2000.

http://www.math.tamu.edu/~dallen/masters/egypt_babylon/babylon.pdf ; *Babylonian Mathematics*

Egyptian Hieroglyphic Numerals and Hieratic Numerals

https://en.wikipedia.org/wiki/Egyptian_numerals; last edited on 18 September 2017, at 20:59.

https://en.wikipedia.org/wiki/Ostracon; last edited on 3 September 2017, at 21:36.

http://www-history.mcs.st-and.ac.uk/HistTopics/Egyptian_numerals.html; Article by: *J J O'Connor* and *E F Robertson* December 2000 "Egyptian numerals"; *Egyptian History Topics Index.*

https://www.easycalculation.com/funny/numerals/egyptian.php; *Egyptian numerals*

https://www.reference.com/history/advantages-egyptian-numerals-c109910635e4fa4e; *What are the advantages of Egyptian numerals?*

http://storyofmathematics.com/egyptian.html; Luke Mastin 2010.

http://www.ancientegyptianfacts.com/ancient-egyptian-numbers.html; "Ancient Egyptian numbers and numeral system". *Ancient Egyptian facts 2017.*

Phoenician Numerals

https://en.wikipedia.org/wiki/Phoenician_alphabet#Numerals; last edited on 10 September 2017, at 06:21.

http://www.anticopedie.fr/mondes/mondes-gb/phenicie-langue.html; "The Phoenicians: language, writing, numbers".

http://scriptsource.org/cms/scripts/page.php?item_id=entry_detail&uid=gh4rjqbhl2

https://phoenicia.jimdo.com/the-phoenicans/

Greek Numerals

https://en.wikipedia.org/wiki/Greek_numerals; last edited on 12 September 2017, at 16:15.

http://www-history.mcs.st-andrews.ac.uk/HistTopics/Greek_numbers.html; "Greek Number System"; Article by: J J O'Connor and E F Robertson; January 2001.

http://www.sciencedirect.com/science/article/pii/0315086082901379; Minoan and Mycenaean numerals, In Memory of Ken May, Author Dirk J Struik, *Historia Mathematica*, Volume 9, Issue 1, February 1982, pp. 54-58.

http://www.unicode.org/charts/PDF/U10100.pdf; Aegean Numbers Range: 10100–1013F This file contains an excerpt from the character code tables and list of character names for The Unicode Standard, Version 10.0; The Unicode Standard 10.0, Copyright © 1991-2017 Unicode, Inc. All rights reserved.

https://en.wikipedia.org/wiki/Aegean_numerals; last edited on 16 September 2017, at 20:33.

https://en.wikipedia.org/wiki/Attic_numerals; last edited on 27 May 2017, at 17:37.

http://dictionary.sensagent.com/Attic%20numerals/en-en/

https://wikivisually.com/wiki/Cypro-Minoan_script.

https://en.wikipedia.org/wiki/Cypro-Minoan_syllabary; last edited on 3 October 2017, at 14:37.

https://en.wikipedia.org/wiki/Cypriot_syllabary; last edited on 25 August 2017, at 04:40.

https://en.wikipedia.org/wiki/Minoan_civilization; last edited on 30 September 2017, at 23:45.

Roman and Etruscan Numerals

https://en.wikipedia.org/wiki/Roman_numerals; last edited on 30 September 2017, at 06:39.

https://en.wikipedia.org/wiki/Old_Italic_script; last edited on 11 August 2017, at 20:46.

https://en.wikipedia.org/wiki/Etruscan_numerals; last edited on 24 September 2017, at 07:54.

https://en.wikipedia.org/wiki/Etruscan_language; last edited on 24 September 2017, at 07:54.

https://en.wikipedia.org/wiki/Old_Italic_script#Etruscan_alphabet; last edited on 11 August 2017, at 20:46.

https://en.wikipedia.org/wiki/Etruscan_civilization; last edited on 25 September 2017, at 05:34.

Amharic Numerals of Ethiopia

http://www.geez.org/Numerals/; A Look at Ethiopic Numerals.

https://en.wikipedia.org/wiki/Amharic; last edited on 12 October 2017, at 14:46.

https://www.learn-amharic.com/ethiopic-numerals; by Adam Young.

https://ethiopia.limbo13.com/index.php/numbers/; Road to Ethiopia: Camino a Etiopia, Numbers in Amharic, June 18, 2008 by Alicia.

http://omniglot.com/language/numbers/amharic.htm; Omniglot; online encyclopedia.

https://www.language-school-teachers.com/rte/rte/LessonPhraseView.asp?Id=450: Amharic numbers: 0 to 10 (Part 1).

http://www.omniglot.com/writing/ethiopic.htm; Ge'ez Script.

http://aboutworldlanguages.com/amharic; Amharic by Irene Thompson, June 28, 2016.

http://www.lexilogos.com/keyboard/amharic.htm; Multilingual keyboard: Amharic.

Hebrew and Arabic Abjad Numerals

https://en.wikipedia.org/wiki/Abjad_numerals; last edited on 22 June 2017, at 17:04.

http://www.omniglot.com/language/numbers/hebrew.htm; *compiled partly by Michelle Abramowitz.*

https://en.wikipedia.org/wiki/Hebrew_numerals; last edited on 2 October 2017, at 23:21.

https://en.wikipedia.org/wiki/Abjad; last edited on 20 June 2017, at 02:37.

https://en.wikipedia.org/wiki/Gematria; last edited on 11 October 2017, at 12:47.

https://en.wikipedia.org/wiki/About_the_Mystery_of_the_Letters; last edited on 11 November 2016, at 15:49.

https://en.wikipedia.org/wiki/Biblical_numerology; last edited on 30 August 2017, at 05:11.

https://en.wikipedia.org/wiki/Numerology; last edited on 16 October 2017, at 01:56.

https://en.wikipedia.org/wiki/Significance_of_numbers_in_Judaism; last edited on 29 September 2017, at 10:25.

https://en.wikipedia.org/wiki/Number_of_the_Beast; last edited on 6 September 2017, at 23:52.

https://en.wikipedia.org/wiki/Numbers_in_Egyptian_mythology; last edited on 14 July 2017, at 21:48.

https://en.wikipedia.org/wiki/Numbers_in_Norse_mythology; last edited on 24 June 2017, at 23:44.

https://en.wikipedia.org/wiki/Kabbalah; last edited on 26 September 2017, at 02:44.

http://web.stanford.edu/dept/lc/arabic/alphabet/; The Arabic Alphabet – Chart, Stanford University, Language Center, Arabic Department, Original site by Khalil Barhoum and Joseph Kautz, current upgrade by Tony Jin.

https://en.wikipedia.org/wiki/Hebrew_alphabet; last edited on 7 October 2017, at 07:16.

https://en.wikipedia.org/wiki/Arabic_alphabet; last edited on 11 October 2017, at 07:40.

https://en.wikipedia.org/wiki/Isopsephy; last edited on 25 August 2017, at 15:52.

Armenian and Georgian Numerals

https://en.wikipedia.org/wiki/Armenian_numerals; last edited on 27 September 2017, at 12:30.

https://en.wikipedia.org/wiki/Georgian_numerals; last edited on 7 June 2017, at 16:28.

Kharosthi, Brahmi and Devanagari Numerals

http://www.scit.wlv.ac.uk/~cm1993/maths/mm2217/han.htm; Hindu-Arabic Numerals; pages maintained by M.I.Woodcock.

https://en.wikipedia.org/wiki/Kharosthi#Numerals; last edited on 10 October 2017, at 09:15.

https://infogalactic.com/info/Kharosthi; last modified on 23 December 2015, at 03:08.

https://en.wikipedia.org/wiki/Brahmi_numerals; last edited on 14 September 2017, at 08:17.

https://en.wikipedia.org/wiki/Brahmi_script; last edited on 18 October 2017, at 06:10.

https://infogalactic.com/info/Brahmi_numerals; last modified on 17 April 2015, at 23:41.

http://www-groups.dcs.st-and.ac.uk/history/HistTopics/Indian_numerals.html; Indian Numerals **by:** *J J O'Connor* and *E F Robertson*, November 2000.

http://www.thefullwiki.org/Devanagari_numerals.

https://pparihar.com/2017/09/20/evolution-of-modern-numerals-bramhi-india/.

https://en.wikipedia.org/wiki/Devanagari; last edited on 16 October 2017, at 01:18.

https://en.wikipedia.org/wiki/Sanskrit; last edited on 17 October 2017, at 06:42.

Gujarati, Gurmukhi, Bengali and Oriya Numerals

https://en.wikipedia.org/wiki/Gujarati_numerals; last edited on 27 April 2017, at 04:16.

https://en.wikipedia.org/wiki/Gurmukhi_numerals; last edited on 12 May 2017, at 23:59.

http://www.omniglot.com/language/numbers/punjabi.htm.

http://www.omniglot.com/writing/punjabi.htm.

https://en.wikipedia.org/wiki/Bengali_language; last edited on 19 October 2017, at 16:10.

http://www.omniglot.com/writing/bengali.htm.

http://www.newworldencyclopedia.org/entry/Bengali_language; last modified on 1 June 2016, at 20:29.

https://en.wikipedia.org/wiki/Odia_language; last edited on 9 October 2017, at 13:44.

https://www.britannica.com/topic/Oriya-language; last updated 7/27/2017.

http://www.omniglot.com/writing/oriya.htm.

Telugu, Kannada, Tamil and Malayalam Numerals

https://en.wikipedia.org/wiki/Telugu_language; last edited on 15 October 2017, at 18:26.

http://www.omniglot.com/writing/telugu.htm.

http://www.omniglot.com/writing/kannada.htm.

https://www.britannica.com/topic/Kannada-language; written by Bhadriraju Krishnamurti, last updated 7/28/2017.

https://en.wikipedia.org/wiki/Kannada; last edited on 18 October 2017, at 13:31.

http://www.omniglot.com/writing/tamil.htm.

https://www.britannica.com/topic/Tamil-language; written by Bhadriraju Krishnamurti, last updated 9/12/2017.

https://en.wikipedia.org/wiki/Tamil_language; last edited on 2 October 2017, at 14:18.

http://www.omniglot.com/writing/malayalam.htm.

https://www.britannica.com/topic/Malayalam-language; written by Bhadriraju Krishnamurti, last updated 7/27/2017.

https://en.wikipedia.org/wiki/Malayalam; last edited on 19 October 2017, at 10:43.

Sinhala Numerals of Sri Lanka

https://en.wikipedia.org/wiki/Sinhala_numerals; last edited on 28 August 2017, at 05:01.

http://www.omniglot.com/language/numbers/sinhala.htm. Information by Zein Al-A'bideen Shabeeb and Ajit Vijēsiṅha Hakmanagē.

http://www.omniglot.com/writing/sinhala.htm.

Javanese Numerals

https://en.wikipedia.org/wiki/Javanese_script; last edited on 1 October 2017, at 14:38.

https://en.wikipedia.org/wiki/Javanese_numerals; last edited on 26 January 2017, at 18:44.

http://www.omniglot.com/writing/javanese.htm.

Burmese Numerals

https://en.wikipedia.org/wiki/Burmese_numerals; last edited on 20 August 2017, at 21:3.

https://en.wikipedia.org/wiki/Burmese_alphabet; last edited on 3 September 2017, at 06:23.

https://en.wikipedia.org/wiki/Burmese_language; last edited on 14 October 2017, at 05:19.

Khmer, Thai and Lao Numerals

https://en.wikipedia.org/wiki/Khmer_language; last edited on 21 October 2017, at 11:49.

https://www.britannica.com/topic/Austroasiatic-languages#ref604358; by Gerard Diffloth.

https://www.britannica.com/topic/Khmer-language.

https://en.wikipedia.org/wiki/Khmer_alphabet; last edited on 23 September 2017, at 11:45.

https://en.wikipedia.org/wiki/Khmer_numerals; last edited on 8 May 2017, at 10:04.

https://en.wikipedia.org/wiki/Thai_numerals; last edited on 13 May 2017, at 19:06.

https://en.wikipedia.org/wiki/Thai_language; last edited on 9 October 2017, at 00:58.

http://www.omniglot.com/writing/thai.htm.

https://infogalactic.com/info/Thai_numerals; last modified on 16 August 2015, at 15:45.

https://en.wikipedia.org/wiki/Lao_alphabet#Numerals; last edited on 11 October 2017, at 19:01.

Tibetan and Mongolian Numerals

https://en.wikipedia.org/wiki/Standard_Tibetan#Numerals; last edited on 26 September 2017, at 23:10.

https://en.wikipedia.org/wiki/Tibetan_people; last edited on 19 October 2017, at 11:33.

https://en.wikipedia.org/wiki/Tibetan_alphabet; last edited on 6 October 2017, at 16:23.

https://infogalactic.com/info/Mongolian_numerals; last modified on 4 January 2014, at 04:03.

https://en.wikibooks.org/wiki/Mongolian/Numerals_and_Fractions; last edited on 11 April 2017, at 07:44.

http://www.answers.com/Q/What_are_the_mongolian_numerals_1_to_20.

http://www.omniglot.com/writing/mongolian.htm.

Numerals used in Arabic (Eastern Arabic Numerals)

https://en.wikipedia.org/wiki/History_of_the_Hindu%E2%80%93Arabic_numeral_system; last edited on 18 September 2017, at 22:23.

https://en.wikipedia.org/wiki/Brahmagupta; last edited on 22 October 2017, at 18:40.

https://en.wikipedia.org/wiki/Muhammad_ibn_Musa_al-Khwarizmi; last edited on 22 October 2017, at 04:56.

https://en.wikipedia.org/wiki/Al-Kindi; last edited on 22 October 2017, at 20:07.

https://en.wikipedia.org/wiki/Abu%27l-Hasan_al-Uqlidisi; last edited on 16 September 2017, at 02:24.

https://en.wikipedia.org/wiki/Arabic_numerals; last edited on 23 October 2017, at 17:36.

Persian Numerals

https://www.britannica.com/topic/Persian-language.

Western Indo-Arabic Numerals Introduced to Europe

https://en.wikipedia.org/wiki/Arabic_numerals; last edited on 23 October 2017, at 17:36.

https://en.wikipedia.org/wiki/Codex_Vigilanus; last edited on 25 March 2017, at 20:53.

www.ingramcontent.com/pod-product-compliance
Lightning Source LLC
Chambersburg PA
CBHW050151230526
45470CB00001B/47
9781979614849